DESIGN OF ENERGY-EFFICIENT APPLICATION-SPECIFIC INSTRUCTION SET PROCESSORS

Design of Energy-Efficient Application-Specific Instruction Set Processors

by

Tilman Glökler

IBM Deutschland Entwicklung GmbH,
Böblingen, Germany

and

Heinrich Meyr

Integrated Signal Processing Systems,
Aachen University of Technology, Germany

KLUWER ACADEMIC PUBLISHERS
BOSTON / DORDRECHT / LONDON

A C.I.P. Catalogue record for this book is available from the Library of Congress.

ISBN 978-1-4419-5425-1 e-ISBN 978-1-4020-2540-2

Published by Kluwer Academic Publishers,
P.O. Box 17, 3300 AA Dordrecht, The Netherlands.

Sold and distributed in North, Central and South America
by Kluwer Academic Publishers,
101 Philip Drive, Norwell, MA 02061, U.S.A.

In all other countries, sold and distributed
by Kluwer Academic Publishers,
P.O. Box 322, 3300 AH Dordrecht, The Netherlands.

Printed on acid-free paper

Softcover reprint of the hardcover 1st edition 2004

Contents

Foreword

It would appear we have reached the limits of what is possible to achieve with computer technology, although one should be careful with such statements – they tend to sound pretty silly in five years.

John von Neumann, 1949.

Application-specific instruction set processors (ASIPs) have the potential to become a key building block of future integrated circuits for digital signal processing. ASIPs combine the flexibility and competitive time-to-market of embedded processors with the computational performance and energy-efficiency of dedicated VLSI hardware implementations. Furthermore, ASIPs can easily be integrated into existing semi-custom design flows: the ASIP designer has full control of the implementation and verification. As ASIPs replace commercial embedded processors, there is no need to pay royalties to third parties.

This book was written for hard- and software design engineers as well as students with a fundamental knowledge of VLSI logic design. The benefits of ASIPs can only be exploited by designers with expertise in the fields of VLSI hardware, computer architecture, and embedded software design. This book provides the essential knowledge in each of these disciplines and focuses on the practical implementation of ASIPs for real-world applications. Many examples illustrate the proposed methodology; theoretic discussions are kept to the minimum.

This book constitutes my Ph.D. thesis, which has been performed at the Institute for Integrated Signal Processing Systems at Aachen University of Technology (ISS/RWTH Aachen/Germany). My reviewers encouraged me to extend my thesis and publish this comprehensive book about ASIP design.

The first chapter of this book introduces the advantages of ASIPs and motivates the requirement for an elaborated design methodology. In Chapter 2, the focus of this work is described in detail and an overview of related work is given. Chapter 3 introduces and summarizes the basics of low-energy VLSI design. This chapter is a prerequisite for the design space definition of ASIPs and the discussion of critical factors for energy-efficient ASIP architectures in Chapter 4. The proposed ASIP design flow is presented in Chapter 5 with a special focus on design tasks to obtain an energy-efficient implementation. The LISA tool suite, which was developed at the ISS, and enhancements of these tools triggered by this work are presented in Chapter 6. The described tools support the generation of critical hardware parts in order to save energy as well as the verification of the implemented ASIP hard- and software. Quantitative results of two case studies are given in Chapter 7, which prove the applicability of the proposed design flow and the developed tools. The first case study demonstrates the impressive potential of ASIP performance and energy optimizations, whereas the second case study compares the architectural and implementation efficiency of two different ASIP design approaches.

Acknowledgments

I would like to express my sincere gratitude to Professor Heinrich Meyr, the coauthor of this book, for supervising my Ph.D. thesis. Frequent discussions with him added greatly to my work and his guidance and support have been invaluable to my academic development over the last five years.

Furthermore, I would like to thank Professor Stefan Heinen for generously spending his time to advise me and for his valuable contribution to improve this thesis.

I am expecially grateful to Dr. Stephan Bitterlich for many fruitful discussions and various good ideas. Moreover, I would like to thank all my colleagues at the Institute for Integrated Signal Processing Systems (ISS) for the pleasant five common years. Special recognition is given to Tim Kogel, Dr. Falco Munsche, Dr. Jens Horstmannshoff, Oliver Wahlen and Manuel Hohenauer for proof-reading and for many valuable proposals to improve this thesis. I also would like to thank Oliver Schliebusch for updating the LISA appendix. Moreover, I am very grateful to Dr. Stefan A. Fechtel and the design team of Infineon Technologies AG for supporting the ICORE DVB-T chip project.

Finally, I would like to thank my parents for their support during my studies and my girlfriend Eva-Marie for her comprehension and patience during the writing of this thesis.

Tilman Glökler
October, 2003

About the Authors

T. Glökler received his diploma degree with honors in Electrical Engineering from Technical University of Stuttgart, Germany, in 1997. He spent five years working on his Ph.D. thesis at the Institute for Integrated Signal Processing Systems (ISS) at Aachen University of Technology (RWTH Aachen). At the ISS he was primarily involved in ASIP design and low-power hardware design methodology as well as in the development of EDA tools. He has written about 10 scientific conference and journal papers. His research interests include advanced algorithms for design automation and digital signal processing with a special focus on programmable architectures for efficient HW/SW codesign. Currently, he is with IBM Deutschland Entwicklung GmbH, Germany, where he is working on the design and verification of high-end microprocessors for consumer applications. Tilman Glökler is a member of the IEEE.

H. Meyr received his M.Sc. and Ph.D. from ETH Zurich, Switzerland. He spent over 12 years in various research and management positions in industry before accepting a professorship in Electrical Engineering at Aachen University of Technology (RWTH Aachen) in 1977. He has worked extensively in the areas of communication theory, digital signal processing and CAD tools for system level design for the last thirty years. His research has been applied to the design of many industrial products. At RWTH Aachen he is a co-director of the institute for integrated signal processing system (ISS) involved in the analysis and design of complex signal processing systems for communication applications. He was a co-founder of CADIS GmbH (acquired 1993 by Synopsys, Mountain View, California) a company which commercialized the tool suite COSSAP. In 2001 he has co-founded LISATek Inc., a company with breakthrough technology to design application specific processors. Recently (February 2003) LISATek has been acquired by CoWare, an acknowledged leader in the area of system level design. At CoWare Dr. Meyr has accepted the position of Chief Scientist. He also serves as a member of the board of directors at CoWare and another large corporation. Dr. Meyr has published numerous IEEE papers and holds many patents . He is author (together with Dr. G. Ascheid) of the book "Synchronization in Digital Communications", Wiley 1990 and of the book "Digital Communication Receivers. Synchronization, Channel Estimation, and Signal Processing" (together with Dr. M. Moeneclaey and Dr. S. Fechtel), Wiley, October 1997. He has received two IEEE best paper awards. Dr.Meyr is also the recipient of the prestigious Vodafone Innovation Prize for the year 2000. The Vodafone prize is awarded for outstanding contribution to the area of wireless communication As well as being a Fellow of the IEEE he has served as Vice President for International Affairs of the IEEE Communications Society.

List of Figures

List of Tables and Examples

Chapter 1

Introduction

In the last decade, integrated digital circuits have emerged as the computational core of many digital devices in everyday life. Examples are mobile phones, organizers, personal computers, networking devices and embedded systems for automotives and industrial automation. The economical importance of digital devices is steadily increasing with an average annual growth rate in semiconductor sales of about 15% since the development of the microprocessor [38].

As Moore's law is expected to be valid for at least the next decade [212], the capability and complexity of digital devices will continue to grow. However, the growth in design productivity for digital circuits cannot keep up with the technological growth [136]. This gap represents a serious bottleneck for the implementation of new competitive devices. Especially for embedded digital circuits, a shift from hardware to software implementations is a solution to this issue. Increasing the software part of a design improves the design productivity due to the simplicity of the software implementation process and due to the increased design reuse factor. Furthermore, this shift to software in embedded systems (embedded software) enables more designers to participate in the implementation process.

Moore's law results in an increasing percentage of software-realizable implementations for all applications with a constant computational complexity[1] due to the exponential increase in available processing power. Embedded software typically has to meet tight real-time constraints and/or high energy-efficiency requirements, especially for mobile appliances. For these two reasons, specialized embedded processors like commercial DSPs or microcontrollers are becoming more and more popular. These embedded processors provide a significantly better cost/performance ratio than general purpose processors for desktop applications. Nevertheless, embedded processors are still limited

[1] This is also true for applications with exponentially increasing computational complexity provided that the exponential increase in computational complexity is smaller than the technological increase.

in terms of computational performance, because they use a fixed non-application-specific instruction set architecture. Furthermore, these processors typically expose poor energy-efficiency compared to more application-specific implementations, because they target a broad range of embedded applications.

Energy-efficiency and flexibility are competing goals for a hardware implementation. Figure 1.1 depicts this tradeoff for several implementation paradigms: embedded standard processors, DSPs, FPGAs and dedicated hardware. The so-called *application-specific instruction set processors* (ASIPs) are able to fill the energy-flexibility gap between dedicated hardware and programmable DSPs for a given application according to Figure 1.1. ASIPs take advantage of user-defined instructions and a user-defined data path optimized for a certain target application. The result of this optimization is a higher computational performance than general purpose approaches and a better energy-efficiency. This is one reason for the current industrial trend to use more and more customized processors [23]. This trend can be explained from the perspective of both hardware and software designers.

From the hardware designers' point of view, ASIPs considerably facilitate the implementation of tasks that require a high degree of flexibility. This flexibility is needed to track evolving standards and for implementations that are prone to late design changes. Furthermore, the design time is decreased especially due to the high reuse factor of software-based implementations. This fact is particularly important for redesigns with the goal to implement distinguishing features in an existing product for competitive reasons. Finally, the ASIP tasks can be modeled with high level languages, which provide a rapid and methodical approach to the design of resource shared hardware. Synthesizable ASIPs are technology-independent and can easily be integrated in any established semi-custom design flow together with other hardware blocks.

From the software point of view, ASIPs offer a new degree of freedom for optimization: The design input for ASIPs is both the software implementation in form of a high level language description as well as the ASIP hardware architecture in form of a hardware description language. The new degree of freedom for software designers, the hardware architecture, removes the traditional upper bound in computational performance of conventional fixed processor architectures by introducing

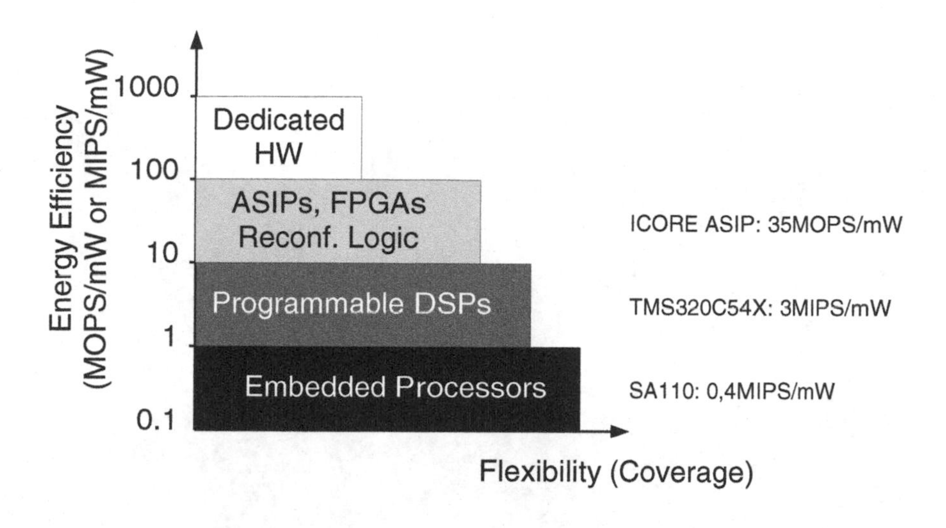

Figure 1.1: The Energy-Flexibility Gap (Source: [1] with modifications)

scalability of processor resources. Therefore, oversized and energy-wasting fixed processor cores can be replaced by energy-efficient ASIPs to meet the performance constraints of an embedded application.

ASIP design is a complex optimization problem requiring expertise in VLSI logic, computer architecture and application software design. The complexity of this design task makes it difficult for the designer to explore a large number of design alternatives in order to find an optimum implementation within a competitive design time. Furthermore, ASIP design for systems with tight energy constraints leads to additional complexity, which aggravates this issue.

This thesis presents a solution to this complexity problem by providing an optimized design methodology for ASIPs considering the typical performance and energy constraints of mobile embedded systems. The feasibility of the proposed design methodology is proven with two case studies. Furthermore, typical ASIP optimizations are introduced and evaluated in order to assess the potential of best practice ASIP implementations over fixed processor architectures.

Chapter 2

Focus and Related Work

This chapter presents the motivation and the focus of this thesis as well as the essential differences to existing approaches. Moreover, an overview of related work concerning ASIP design for low-energy consumption is given.

2.1 Focus of This Work

The focus of this work are application-specific instruction set processors (ASIPs) for embedded DSP applications with performance and energy constraints. Energy in this context refers to the energy that is consumed for a given well-defined computational task. This metric corresponds to the average power that is consumed for the same task.

The proposed methodology primarily targets (but is not limited to) semi-custom designs. This design approach enables the use of a high level of abstraction for design entry, whereas the degrees of freedom for optimization are moderately decreased compared to full-custom design due to the constraints imposed by the standard cell library supplier.

This work has the goal to answer the following scientific problems:

- How much can be gained in performance and energy-efficiency using ASIPs instead of general purpose processors?
- To which extend can energy-driven ASIP optimizations increase the energy-efficiency?
- How can ASIPs be designed in order to meet the performance, energy and/or area constraints?
- Does the proposed design methodology enable a competitive time-to-market?

In order to answer these questions, several case studies have been performed and extensive optimization techniques have been developed and evaluated. These optimization techniques include general low-power optimizations for dedicated hardware as well as ASIP-specific techniques.

Furthermore, tool-based methodologies for the instruction encoding and for the generation of energy-critical ASIP parts as well as enhanced verification techniques have been developed. This is especially important to obtain a competitive design time for the ASIP implementation.

2.2 Previous Work

At present, there is no publication covering a design methodology for ASIPs that enables to jointly optimize the performance, the silicon area and the power consumption. This fact is also emphasized by Jain [132]. The published work rather focuses on performance optimization, some publications also cover the tradeoff between performance and silicon area.

Publications related to low-power ASIP design can be subdivided into the topics **ASIP design methodologies**, **ASIP case studies**, and **basic low-power design techniques** for general purpose processors and dedicated hardware. Furthermore, **ASIP verification** (which is a subtopic of ASIP design methodologies) is of paramount importance to obtain working silicon and represents a tedious and time-consuming design task. The following subsections provide an overview of literature covering these four topics.

2.2.1 ASIP Design Methodologies

This summary of ASIP design methodologies does not discuss in detail general purpose processor designs and significantly incomplete ASIP design environments without a path to hardware implementation like BUILDABONG [242] PARTITA [50], ISPS [22], [34], the work of Engel [69] and of Bajot [20]. Furthermore, methodologies with a significantly incomplete software design tool chain like READ [145] as well as various compiler-centric publications on ASIP code generation [102] [103] [52] [99] [130] [165] [166] [170] [208] [230] [247] [254] are not explicitly covered.

Existing ASIP design environments can be differentiated according to the flexibility to support various processor classes. Many design environments use predefined, largely invariant processor templates and software design tools, covering a limited ASIP design space. Other environments provide generic processor description languages, which enable the designer to add user-defined structures to an existing processor or to describe entirely new processor architectures, often at the expense of the quality of the available software design tools.

Commercial approaches targeting largely fixed processor templates include the Xtensa core of Tensilica [243] [97], the ManArray architecture of BOPS [31], the ARCtangent processor of ARC [12] [13], the Jazz processor of Improv [261] [160] and the R.E.A.L. DSP of Philips [141]. Further work on largely fixed processor classes include Flexware [202], the research project PICO [211] [2], Satsuki [225], and ASIA [116] [117].

Xtensa is a moderately parameterizable RISC (reduced instruction set computer) load/store architecture with variable length instructions (24 or 16 bit), 3 operand instructions, and about 80 base instructions. Parameters of the processor comprise the choice of a 32 or 64 general purpose register file, the size of caches, the write buffer size, the endianess and the availability of certain instructions like multiply-accumulate etc. The automatically generated design tools for each specific processor instance of Xtensa include a C compiler, assembler, linker and debugger. Furthermore, the user can define new instructions and additional functional units using the TIE[1] language.

The ManArray architecture of BOPS uses the concept of a multi-processor system, which is optimized for DSP applications like wireless applications, multimedia and image processing. Each of the processor elements is a RISC core with a fixed so-called indirect VLIW[2] (iVLIW) architecture, which is implemented by a VLIW-look-up-table and spe-

[1] Tensilica Instruction Extension

[2] VLIW is the abbreviation for very long instruction word.

cial 32 bit instructions to execute one of the stored VLIW instructions[3]. The instruction set supports typical DSP instruction, subword-level parallelism, and also typical micro-controller features like bit manipulation and low-latency interrupts. The focus of the ManArray architecture is the scalability of a parameterizable number of tightly interconnected processing elements for regular algorithms requiring high computational performance.

The ARCtangent-A4 microprocessor of ARC is a unified RISC/DSP core with a 4 stage pipeline architecture and configurable functional units, memories and an extensible instruction set. The core is delivered as a soft-core together with software development tools including a DSP function library.

The Jazz processor of Improv is a customizable building block embedded into a generic platform (PSA - programmable system architecture) comprising several typically different instances of the processor together with data/instruction memory and I/O blocks. The Jazz processor represents a memory-register VLIW-architecture using a set of predefined computational units in combination with high bandwidth to data memory. For the specific instances the user can configure parameters like the data width, the depth of the hardware task queue and the number and kind of computational units within certain constraints. For the selected architecture configuration, software design tools as well as synthesizable HDL code can be automatically generated.

The R.E.A.L. DSP of Philips uses a customizable base architecture with 2 multipliers and 4 ALUs in combination with a general purpose register file. Instruction formats with 16 and 32 bits as well as so-called ASI (Application-Specific Instructions), which allow up to 256 VLIW instructions stored in an internally customizable look-up table[4] are supported. The DSP programmer or the high level language (HLL) compiler has to specify the part of the code for parallel execution. The ASI look-up table can be implemented using a RAM for prototypes or a ROM/synthesized block for the processor in the final product.

The Flexware environment is an ASIP design environment based on a simple parameterizable processor template. Configuration parame-

[3]This concept generalizes the idea of the CLIW (configurable long instruction word) architecture of CARMEL (Infineon Technologies [232])

[4]This approach is similar to the above-mentioned iVLIW concept.

ters include the bit width, the number of registers and the number of ALUs as well as the definition of new instructions. The environment provides the typical software design tool chain including the code generator CodeSyn [201] and a hardware description generation back end. Simulation is performed using the VHDL model of the target processor.

PICO as well as Trimaran [255] are both part of the compiler and architecture research group of HP Labs. PICO is an environment that automatically explores the design space for a heterogeneous processor-coprocessor system for applications written in C code. A synthesizable VHDL description for non-programmable processors as well as an optimally configured instance of a VLIW processor called HPL-PD [137] is generated. The approach is limited to this processor type with a fixed instruction set, but it supports different memory and cache configurations.

Satsuki is a design environment, which uses a moderately parameterizable processor template as target architecture. Parameters of this template are data path width, number of registers and instruction and data memory size. Furthermore, a C compiler for single precision integer arithmetic is supported.

ASIA is a system that synthesizes instruction set architectures for a given application, which is available in form of a micro-operation program and a coarse pipeline stage structure of the target architecture. The results of ASIA is the microarchitecture definition of an architecture, which is able to satisfy a given runtime constraint and which uses a data stationary control model,

Commercial ASIP design environments supporting more flexible target architectures are currently being designed by TargetCompiler Technologies [240] and by STMicrolectronics (Flexware2 [199]).

The environment of TargetCompiler Technologies is based on the high level modeling language nML [84]. The supported retargetable tools include a C compiler, an instruction set simulator, an assembler and linker as well as a hardware description generator. The description language nML has been extended to support pipelined architectures. Unfortunately, there is no list of restrictions available, which describe the limitations of the supported processor architecture classes.

Flexware2 [199] is the successor of Flexware and is based on the instruction set description language IDL [200]. The Flexware2 environment enables the generation of instruction set simulator, assembler, and linker. The HLL compiler is based on the COSY framework [3] and needs a separate processor description. Hardware generation from IDL is currently not supported.

Scientific ASIP environments targeting flexible processor architectures comprise PEAS-III [146] [126] and MetaCore [280].

The ASIP design environment PEAS-III uses a textual micro-operation description and provides a GUI for ASIP modeling. PEAS-III enables the generation of a synthesizable hardware model [126] and development tools like a C Compiler [74]. Unfortunately, there is little information available about the supported processor classes and about the quality of results.

The environment MetaCore uses a predefined parameterizable DSP microarchitecture supporting an essential set of basic instructions as well as user-selectable predefined instructions. Furthermore, the designer can add application-specific instructions. The specification of the target processor is achieved using a structural (MSL) and a behavioral (MBL) description language. The generated development tools comprise the entire set of typical software design tools including a GCC-based C compiler, instruction set simulator and profiling tools.

2.2.2 ASIP Case Studies

Over the past few years many ASIPs have been designed in industry and in academia. Table 2.1 provides an overview of relevant academic and industrial ASIP designs and case studies. It has to be emphasized that most of the published ASIP case studies focus on performance rather than power optimization. An exception is the work of Kuulusa [152], which evaluates the effect of instruction set modifications on the power consumption, but without using more extensive architectural optimizations.

Authors, Affiliation and Reference	Design Description/Focus
Kuulusa, Tampere Univ. [152]	configurable DSP core for GSM
B. Kienhuis et al., Philips Research [140] [162]	stream-based data flow arch.
A. Alomary Univ. of Tokyo [8]	GCC-based [229] primitive operations
J. Van Praet, IMEC, Leuven [262]	autom. analysis of instruction bundles, then incremental arch. optimization
A. Fauth, Univ. of Berlin [79][78]	user specified archicture using nML [84] with HW generation
Ing-Jer Huang, USC [115]	focus on instruction definition for a given architecture
F. Onion, UC Irvine [195]	compiler assisted insertion of instructions using chained ops.
Q. Zhao, Eindhoven Univ. [282]	static resource model for high-level ISA and compiler design
J. Gong, UC Irvine [96] [95]	parameterizable VLIW architecture
M. Arnold, Delft Univ. [17]	limited interconnections are investigated
K. Kücükcakar, Escalade Corp. [150]	only incr. modifications of largely fixed arch.
H. Choi, Korea Inst. of Sc. and Tech. [49]	case study with PARTITA-design environment
J.-H. Yang, Korea Inst. of Sc. and Tech. [70]	case study with MetaCore environment
P. Faraboschi M. Gschwind IBM Research Center [101]	multi-cluster VLIW, case study for Prolog and vector prefetching
M. Itoh, Y. Takeuchi, Osaka Univ. [127]	HDL generation from a μ-op. descr. (PEAS-III)
R. Camposano, J. Wildberg GMD [41]	case study with CASTLE environment

Table 2.1: ASIP Case Studies

2.2.3 Basic Low-Power Design Techniques

In the following, several low-power design approaches are discussed. These approaches are applicable to general purpose and application-specific processors, but also to dedicated hardware designs[5]. Due to the fact that a detailed discussion of low-power hardware design techniques is presented in Chapter 3, only the most important approaches are summarized at this point.

There is a variety of publications concerning low-power hardware design in general e.g. [177] [44] [45] [66] [104] [221] [251]. Many of them cover technological issues and also typical full-custom techniques like voltage scaling in combination with parallelization, which are not directly applicable to semi-custom chip design. Other publications also cover algorithmic optimizations like optimized filter coefficients, arithmetic operator minimization, and optimization of number representations. High-level optimizations include scheduling techniques in order to exploit data correlation. Standard circuit optimizations like guarding techniques (which includes clock gating) and precomputation are also treated. A more detailed review of work on this topic is given in Section 3.3.

Many other publications [273] [25] [83] exploit the statistical properties of the encoded values by using redundant additional information or by optimized non-redundant code assignments. The goal of these encoding techniques is to lower the toggle frequency of heavily loaded nodes in order to save power. In [264], an approach is presented to reduce the power in a cache memory-based on physical modifications of the memory architecture, which avoid the decharge activity of the high capacitance bit lines. These approaches are related to the idea that has been used in Section 6.3.1 to reduce the energy consumption in embedded instruction memories.

The following publications focus on architectural and/or instruction set modifications to decrease the power consumption.

In [131] the number of general purpose registers of an ARM7TDMI [5] is varied in order to evaluate its effect on power consumption and runtime. Unfortunately, the power model of this work neglects the reduced energy consumption of the changed register file and only takes into account the energy of the memory accesses.

[5] For the greater part, only those approaches have been selected which are applicable to semi-custom ASIP design.

In [231] an architecture tuning methodology is described that uses a fixed instruction set and tunes the implementation e.g. by adding specialized registers for frequently addressed memory locations etc. The paper provides an evaluation of sample modifications with a limited scope resulting in small power savings.

In [272] an energy-conscious methodology for the exploration of a processor-accelerator system using an ARM-compatible core and a custom accelerator is described. The processor instruction set in this case, however, is not application-specific. Similar work has been performed in [100].

Several ideas concerning instruction sets with the option to generate software for low-power consumption are presented in [18]. One idea is the concept of programmable bypass and forwarding registers, where the compiler decides whether a bypass or a forwarding register can be used instead of a real register as data source with the goal to avoid general purpose register accesses. This concept can be regarded as exposing the microarchitecture to the SW interface, which is not unusual for many VLIW architectures. A similar approach has been adopted for the scalable processor architecture in Section 7.2 of this thesis.

The following publications cover general purpose processor design techniques. Many of the presented ideas are also applicable to semi-custom ASIP design.

In [36] a summary of energy and power metrics for general purpose processor systems is presented together with basic design optimizations in order to increase efficiency. These optimization techniques include voltage scaling[6] and optimum instruction set design including the number of registers and the number and kind of functional units and supported addressing modes. Furthermore, energy-efficient cache design and energy-aware operating systems are discussed.

The publications of Tiwari [249] [250] cover a wide range of optimizations for high performance processors including technological optimizations (low-power libraries), circuit techniques (transistor sizing, logic optimization on register transfer level) and operating system power management techniques. Tiwari also introduces a power model

[6] Voltage scaling is typically not applicable to semi-custom designs due to the lack of characterized standard cells.

[253] for software optimizations, which results in low-power code generations strategies [252] e.g. reduction of memory accesses, energy driven instruction selection, instruction reordering, instruction packing, operand swapping and SW power management.

A microcontroller explicitly optimized for high energy-efficiency is the M•Core of Motorola [187]. Various publications [223] [158] [224] describe the low-power techniques that have been applied for this processor core including selective power-down mechanisms, high code density, rich register set, multiple data sizes support, loop cache, and branch folding. The publications partially include power evaluations of these optimizations. Application-specific adaptations have not been performed for this core since it mainly targets general purpose microcontroller applications.

2.2.4 Verification

The verification task targeted by the tool in Section 6.3.2 of this thesis checks the correctness of the ASIP HW description with respect to a cycle-true instruction set reference model. Even in the case of a automatically generated hardware description, this verification is important to reduce the design risk due to errors in the hardware generation tool.

Theoretically, this verification task can be realized with a formal verification approach like in [11], [135], [167], and [277], but this requires a formal specification of the processor description. In the case of the LISA environment an appropriate formal description is not available. Therefore, functional simulation has to be used, which is also applied by many industrial processor design teams [260] [68] [174].

The task of providing suitable stimuli for this functional simulation is tedious and time consuming, but can be partially automated [43] [149]. Krüger [149] presents a tool which is targeted for self test program generation using a structural description of the processor as input to the test program generator. Chandra [43] discusses a methodology to generate stimuli for the IBM S/390 processor. Chandra's methodology includes techniques like symbolic execution and constraint solving in order to cover boundary conditions. The described methodology is targeted and optimized for a single general purpose processor.

2.3 Differences to Previous Work

As energy consumption is getting more and more important for digital chip designs, low-energy **ASIP design methodologies** are of special scientific interest. All of the previously published ASIP design environments primarily focus on performance optimization. Some of them are also able to evaluate the area consumption and enable performance-area tradeoffs. None of them allows explicit energy optimizations, which can be performed with the design methodology as proposed in this thesis.

A similar statement can be made about the related **ASIP case studies**: none of them systematically evaluates the primary sources of energy consumption and none of them proposes or performs explicit energy optimizations.

The differentiation against the **basic low-power design techniques** mentioned above is the fact, that this thesis focuses on the special case of ASIP-typical energy optimizations by exploiting the large ASIP design space including the user-defined instruction set. Previously published energy optimization techniques are used, but also novel ASIP-specific energy optimizations are developed. In contrast to the related work, thorough evaluations of these optimizations are performed using precise gate-level estimations. These techniques have provided essential concepts for the enhancement of the LISA[7] processor design tools in order to facilitate future low-energy ASIP development.

Verification is an important subtask of complete ASIP design flows in order to guarantee a fully functional implementation. This topic is treated in this thesis by explicitly covering a tool that has been developed for this purpose. The proposed semi-automatic test case generator, which is described in Section 6.3.2, uses a similar approach as Krüger [149]. However, instead of a structural processor description, the behavioral LISA description is used. Furthermore, speculative execution on an instruction set simulator together with user-defined rules guarantee that meaningful test scenarios can be generated in a short amount of time.

[7] Please refer to Chapter 6 for a description of the LISA tools suite.

Chapter 3

Efficient Low-Power Hardware Design

From a hardware perspective, an ASIP represents a complex finite state machine where the state transitions are triggered by the input data and the ASIP software. Consequently, all the techniques for efficient low-power or low-energy hardware design are also applicable to ASIPs. This fact is emphasized in the case of application-specific hardware accelerators that are tightly coupled to the ASIP core to increase the energy-efficiency of the implementation. This approach implicitly includes the ASIP software as being an integral part of the hardware implementation, which has to be optimized with equal effort. However, plain software optimization techniques are beyond the scope of this chapter. The basics of software optimization are briefly treated in in Section 5.3.3.

This chapter defines the critical issues of hardware design that have a major influence on the final result. The basics of low-power CMOS hardware design are described including physical effects, metrics to evaluate different architectures and power estimation techniques. Finally, specific design techniques to increase the energy-efficiency of synthesized semi-custom hardware are presented together with references to practical applications. This chapter represents a prerequisite for Chapter 5 where the ASIP typical hardware and software design issues are treated.

3.1 Metrics of the Implementation and the Hardware Design Methodology

In the following two subsections the terms *architectural efficiency* and *design efficiency* are defined similar to [263] as principal metrics for the quantitative evaluation of different design alternatives. This discussion refers to hardware design in general as well as ASIP development.

3.1.1 Characteristics of the Implementation

Implementation constraints are qualitative or quantitative boundary conditions that have to be fulfilled in order to obtain a feasible implementation for a given signal processing application.

Two classes of constraints can be identified:

- **Precise constraints** have to be fulfilled accurately, which means that even a small deviation between constraint and considered parameter leads to device failure. This type of constraint is obviously only applicable to qualitative or discrete parameters.
- **Minimum (maximum) constraints** are typically met by larger (smaller) or equal values of the considered implementation parameter. Safety margins of the constraints have to be provided as a guard against estimation errors. The quantitative difference between constraint and considered parameter is referred to as "slack". Two subtypes of min./max. constraints can be identified in the case of a feasible implementation with slack$\geq$ 0:
  - magnitude of slack is unimportant
  - magnitude of slack enhances the implementation and has to be optimized

In the following, a list of requirements for the ASIP hard- and software implementation is given. This discussion uses a black box abstraction that conceptually contains the complete ASIP hard- and software implementation. The following qualitative or quantitative parameters apply to this model:

- correct **functionality** of the implementation with respect to the specified bit-true behavior (precisely constrained)
- correct **timing** of interfaces (typ. with min./max. constraints)

- **performance constraints** like computational throughput, bit error rate, acquisition probability, number of processed data packets per time unit (with min./max. constraints and a slack that need to be optimized in order to enhance the performance of the digital system e.g. to obtain a competitive advantage)
- average and peak **power consumption** of the module (both with max. constraints, the average power consumption needs to be minimized e.g. for mobile appliances or to reduce the costs of packages and (active) heatsinks)
- **silicon area** of the module (with max. constraint and a slack that needs to be optimized to reduce fabrication costs)
- **observability and controllability** during operation might be needed to discover functional errors in the implementation or the specification (sometimes with a min. constraint)
- the **routability** of the physical design is determined by the interconnection structures of the design and affects the area utilization (silicon area overhead) as well as the timing closure (not explicitly constrained by most synthesis tools [194] [215] [60])
- **testability and self-testability** is an important feature for consumer products to identify chips with fabrication faults at a very early stage (before bonding) to reduce overall costs (typical with a min. constraint and a slack that has to be optimized for higher fault coverage)
- **flexibility** of an implementation is needed to adapt an implementation to different applications or evolving standards or to fix design errors of the implementation (sometimes with a min. constraint and a slack that needs optimization)
- **reusability** or IP-reusability refers to the degree of genericity and flexibility of a design, which enables the reuse of the same design with minor modifications for similar applications (typ. with an implicit min. constraint, the reusability needs to be increased in order to decrease the design effort for similar applications)

The following characteristics apply to the internal ASIP structure and complement this list of characteristics:

- the **application class** (e.g. digital filtering, speech coding, image transformations etc.) which can be efficiently mapped to the processor data-path is strongly correlated with the above-mentioned flexibility of the implementation (extensions of this application class increase the flexibility of the ASIP)
- *either* **simplicity** of the instruction set architecture is required to enable hand programmability, if no compiler for the architecture is available (this parameter strongly affects the software design time)
- *or* the **instruction set class** (ref. to Chapter 4 for a classification) should be selected in order to use an available compiler design environment e.g. COSY [3], if the ASIP is intended for high level language programming support (a good fit of the instruction set class results in a lower effort for compiler retargeting)

For many of the above-mentioned parameters of the implementation the associated slack between constraint and parameter can be quantitatively evaluated. For a selected set of N important slack values that are subject to explicit optimization, it is useful to define a quantitative efficiency in order to compare architectural alternatives. These slack values S_n can be associated with application-specific weights w_n such that the application-specific *architectural efficiency* η_{arch} for these considered slack values can be defined as follows:

$$\eta_{arch} = \prod_{n=1}^{n=N} \frac{1}{S_n{}^{w_n}} \tag{3.1}$$

The well-known classical efficiency for VLSI circuits $\eta = \frac{1}{AT}$ is a special case of the above mentioned architectural efficiency, which considers the equally weighted parameters silicon area and computational performance (critical path) of an implementation.

3.1.2 Characteristics of the Design Methodology

An ideal design methodology achieves the highest possible architectural efficiency (ref. to Subsection 3.1.1) in zero design time. For practical reasons, a feasible trade-off between these parameters has to be selected. The following characteristics of a design methodology are the degrees of freedom to control this trade-off during the design phase:

- The **modeling style** for a given design task has important effects on the design time and the ability to reuse and verify a design.
  - The **level of abstraction** has to be reasonably selected in a hierarchically organized design to reduce the amount of irrelevant details for the current design task – this hierarchical organization can be viewed as a *vertical partitioning* of design tasks.
  - **Modularization** or *horizontal partitioning* of design tasks, on the other hand, reduces the design time in combination with concurrent engineering (see below).
- **Design automation** in combination with abstraction and appropriate tool support both for design and verification enables to reduce the risk of design errors and to speed up the design process.
- **Debugging** on all levels of abstraction should be facilitated by a transparent design methodology and appropriate modeling styles.
- **Design reuse** is also a means to reduce design time. Design reuse has to be performed wherever it is possible to take advantage of encapsulated, verified modules enabling higher abstraction levels [216].
- **Process organization** is the mapping of required design tasks to the available human resources. A vertical *specialization* of work according to the level of abstraction in the design flow can be used together with overlapping execution to reduce the risk of design flaws and to parallelize and speed up the design flow. On the other hand, horizontal partitioning of the work results in reduced design time due to concurrent engineering as well. Typically, a combination of these two approaches are used depending on the complexity of each design task. However, too fine granular partitioning of design tasks can lead to inefficiencies due to an overhead in communication.

- **Monitoring** of the design process and **project management** is mandatory to adaptively control the process organization in order to meet the deadlines and to identify problems at an early stage.

The *design efficiency* η_{design} can be defined using the architectural efficiency η_{arch} of Subsection 3.1.1 together with the overall design effort T_{design} (in man months) and the weight factor w_{design} as follows:

$$\eta_{design} = \left(\frac{1}{T_{design}}\right)^{w_{design}} \eta_{arch} \tag{3.2}$$

The design efficiency η_{design} can be used to compare different design approaches and methodologies and detect problems a-posteriori in the design flow. However, as mentioned before, the number of different parameters that affect the design efficiency is large and for practical analysis of design issues, a more thorough investigation of the design flow and the application is needed.

A different approach to evaluate different design methodologies and implementations could focus on monetary costs of chip design and chip production. This evaluation model could be easily set up using the costs for designing, prototyping as well as the production costs per chip and the market volume. A more thorough investigation of design costs is given in [121], which focuses on the design of dedicated hardware.

3.2 Basics of Low-Energy Hardware Design

VLSI design for reducing the energy consumption of a device basically involves two different issues: estimation of power consumption and techniques to reduce the power or energy consumption. For two different reasons, a reduction of power or energy consumption is beneficial: reduction of the peak power is needed to avoid problems with voltage drops and ground bouncing within a chip. On the other hand, reduction of the average power is mostly driven by mobile systems to increase the battery lifetime, but also for consumer applications, where device costs due to expensive packages, heat sinks and power supplies are of significance. Furthermore, environmental concerns have triggered low-power initiatives like the *Energy Star* program [259]. Finally, the reliability of a system is increased by lowering the average power due to reduced thermal stress and reduced electromigration [191].

In the following subsections the sources for digital CMOS energy consumption, the energy estimation approaches and techniques to reduce the energy and power consumption are discussed.

3.2.1 Sources of CMOS Energy Consumption

The total power consumption P_{total} of a CMOS circuit with the supply voltage V_{dd} can be described by the following equation:

$$\begin{aligned} P_{total} &= I_{standby}V_{dd} + I_{leakage}V_{dd} + I_{sc}V_{dd} + \alpha_{avg}C_l V_{dd}^2 f_{clk} \quad (3.3) \\ &= P_{standby} + P_{leakage} + P_{short\_circuit} + P_{capacitive} \end{aligned}$$

The **standby current** $I_{standby}$ is typically completely avoided by a proper CMOS circuit style and can usually be neglected. However for certain circuit styles (pseudo NMOS, NMOS pass transistor logic, and memory cores) $I_{standby}$ can be an issue [197].

The **leakage current** $I_{leakage}$ is due to the reverse bias current in the parasitic diodes of the diffusion zones and the bulk region of the MOS transistors and also due to the subthreshold current in the case of gate voltages below the threshold voltage. This effect is an issue in the case of current and future technologies with significantly reduced power supply voltages.

The **short-circuit power** $P_{short\_circuit}$ is due to the hot path in typical CMOS circuits (like the inverter in Figure 3.1) when both transistors are on for a short period of time during transitions. This term depends on the input rise (fall) time (slew rate), the output load and the transistor sizes (internal capacitances and gain factors).

The **capacitive power** $P_{capacitive}$ depends on the average switching probability α_{avg}, the clock frequency f_{clk} and the switched capacitance

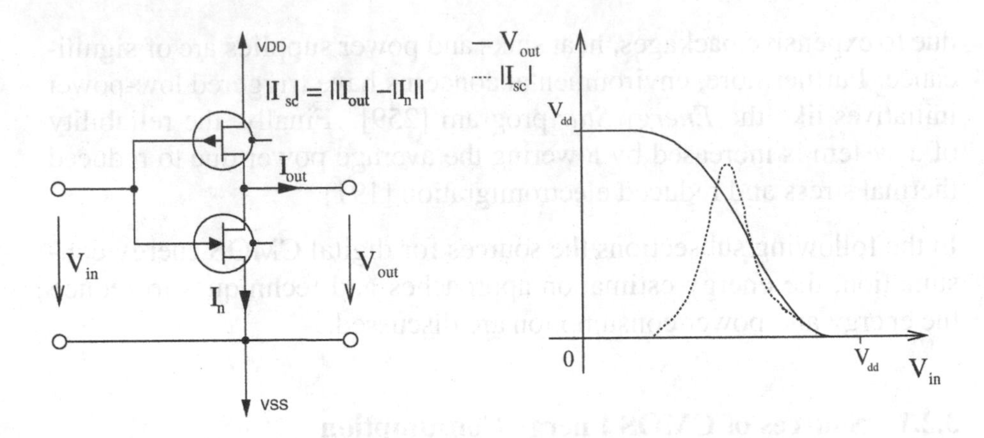

Figure 3.1: Short-Circuit Current in a CMOS Inverter

C_l. It is caused by the power needed to charge and discharge (in most cases parasitic) capacitances on the chip. The switching probability α_{node} (also referred to as toggle activity or as toggle probability) of a single node is defined as the ratio of the number of transitions of the considered logic node to the number of clock transitions within the simulation interval. For strictly synchronous design style using positive (negative) clock edge triggered flip-flops, a logic node transition can only occur after the rising (falling) edge of the clock, thus, the maximum transition probability for a logic node without glitches is $\alpha = \frac{1}{2}$. The transition probability of the clock itself is $\alpha = 1$. The capacitance C_l in Equation 3.4 is the sum of all node capacitances in the considered circuit whereas the α_{avg} is the equivalent average transition probability of C_l according to the following equation:

$$\alpha_{avg} = \frac{\sum_{i=1}^{i=N} \alpha_{node,i} C_{node,i}}{\sum_{i=1}^{i=N} C_{node,i}} = \frac{\sum_{i=1}^{i=N} \alpha_{node,i} C_{node,i}}{C_l} \tag{3.4}$$

where $\alpha_{node,i}$ and $C_{node,i}$ are the transition probabilities and the capacitances respectively of node i.

The term **static power consumption** refers to the sum of $P_{standby}$ and $P_{leakage}$ which typically represent negligible constants for the majority of current CMOS designs.

The sum of $P_{short\_circuit}$ and $P_{capacitive}$ is referred to as **dynamic power** which represents the most significant part of the overall power budget for state-of-the-art CMOS designs. For practical purposes it is useful to replace the short-circuit current I_{sc} with an equivalent short-circuit capacitance $C_{node,sc}$, because the amount of short-circuit power $P_{short\_circuit}$ is also proportional to the switching activity of the circuit nodes. If $C_{node,sc}$ is chosen according to the following equation

$$C_{node,sc} = \frac{I_{node,sc}}{\alpha_{node} V_{dd} f_{CLK}} \tag{3.5}$$

then the effective dynamic capacitance $C_{dyn\_node,i}$ of a node i is can be expressed as follows:

$$C_{dyn\_node,i} = C_{node,i} + C_{node,sc,i} \tag{3.6}$$

Together with

$$C_{dyn} = \sum_{i=1}^{i=N} C_{dyn\_node,i} \tag{3.7}$$

and

$$\alpha_{avg\_dyn} = \frac{\sum_{i=1}^{i=N} \alpha_{node,i} C_{dyn\_node,i}}{C_{dyn}} \tag{3.8}$$

Equation 3.4 can be simplified to

$$P_{total} = I_{standby}V_{dd} + I_{leakage}V_{dd} + \alpha_{avg\_dyn}C_{dyn}V_{dd}^2 f_{clk} \tag{3.9}$$
$$= P_{standby} + P_{leakage} + P_{dynamic}$$

which states more clearly the significant physical effect of the average dynamic toggle probability α_{avg_dyn} on power consumption.

3.2.2 Basic Principles of Lowering the Power Consumption

The following considerations directly refer to the total CMOS power consumption as a goal for minimization. As mentioned before, for current CMOS technologies the major part of CMOS power in Equation 3.10 is the dynamic power consumption:

$$P_{dynamic} = \alpha_{avg\_dyn}C_{dyn}V_{dd}^2 f_{clk} \tag{3.10}$$

The dynamic power consumption can obviously be decreased by

- reducing the supply voltage V_{dd} which results in a quadratic decrease of $P_{dynamic}$
- reducing the clock frequency f_{clk}
- reducing the effective switched capacitance C_{dyn} (which includes the physical node capacitance and the equivalent short-circuit capacitance)
- reducing the switching probability α_{avg_dyn}.

A reduction of the supply voltage V_{dd} increases the combinational circuit delay T_{prop} (not the interconnection delay) according to

$$T_{prop} \propto \frac{V_{dd}}{(V_{dd} - V_t)^{\alpha}} \quad \text{with} \quad \alpha \in [1.0, 2.0] \tag{3.11}$$

which has to be compensated for systems with high throughput e.g. by using parallelized or pipelined processing units. Unfortunately, replication results in a higher power consumption, too, but the quadratic decrease of power consumption due to voltage scaling outweighs this increase in many practical cases like in [45]. A different approach is presented in [71], where the design is partitioned into critical regions with a small timing slack and uncritical regions with a high timing slack respectively using a dual supply voltage without affecting the critical path. For very short feature sizes the exponent α in Equation 3.11 approaches 1.0 and for $V_{dd} >> V_t$ the delay is nearly a constant, which is very favorable for voltage scaling.

However, voltage scaling is limited by technological parameters such as subthreshold leakage current, leakage power and reliability issues due to signal integrity [204] [72].

A reduction of the clock frequency f_{clk} without further changes obviously results in a proportional reduction of the computational performance for synchronous circuits as well as in a proportional reduction of power consumption. This reduction is limited by the minimum computational performance required by an application. The total energy to perform a given computational task is unaffected by a reduction of f_{clk}, if no additional changes are applied.

A reduction of the effective switched capacitance C_{dyn} potentially results in a higher computational performance because of reduced interconnection and transition delays. Typically, this reduction has to be achieved using different hardware architectures or technologies. However, the minimization of C_{dyn} is limited by architectural bounds (due to the minimum required interconnection structure) and technological bounds (due to high interconnection capacitances and unavoidable short-circuit currents). Nevertheless, the logic designer can reduce C_{dyn} on the architectural level by using local instead of global interconnections e.g. with systolic arrays or clustered arithmetic using segmented communication buses.

A reduction of the toggle activity α_{avg_dyn} is typically also achieved with optimized hardware architectures. This reduction is especially effective if applied to logic nodes with a high node capacitance $C_{dyn\_node,i}$, e.g. highly loaded chip pads, interconnection buses etc.

However, this reduction is limited by information theoretical bounds (due to the minimum needed communication resulting in toggle activity as a function of the signal entropies [258] [226]). It is actually possible to exploit data redundancy (e.g. data correlation, non-uniform data distribution etc.) to reduce the toggle activity on certain nodes.

For power and energy critical applications like embedded μ-processors all of the above mentioned parameters are optimized for state of the art devices as described in [14]. However, for semi-custom devices the pressure for low-power design techniques is obviously not yet as critical. Typically, the supply voltage for semi- custom technologies is restricted to a certain range $[V_{min}, V_{max}]$. Only for predefined working conditions, which include defined sets of values for the supply voltage, the operating temperature and the quality of the fabrication process, the relevant electrical parameters, the power consumption and the delay for cells and interconnection of the target technology are available. Beyond these working conditions, correct functionality of the chip is not guaranteed by the foundry. This makes it impossible for a conservative designer to use voltage scaling beyond V_{min}. Nevertheless, for aggressive low-power applications, the usable voltage range might be extended below V_{min}.

For semi-custom design, the remaining degrees of freedom, namely clock frequency, switching activity and capacitance reduction have to be simultaneously optimized to maximize the power savings.

3.2.3 Measuring and Quantifying Energy-Efficiency

In order to obtain precise values of the power or energy consumption, appropriate analysis techniques are necessary. Power analysis techniques for semi-custom chips can take advantage of the different levels of abstractions, namely, the circuit level, the cell and the RTL level. The layout extraction of each standard cell performed by the technology vendor results in an equivalent schematic using resistors, capacitors, inductors and current/voltage sources. This schematic or parameterized model can then be simulated with SPICE [182] or similar analysis tools using *transient analysis* to obtain precise estimates for the switching power of the cell. These simulations have to be performed for all the defined working conditions and are often calibrated with measurement data from actual test chips. After this process, the simulated and measured values can be used as library data for power analysis steps at the cell level.

At the cell level, so-called *gate-level simulations* of the synthesized netlist of standard cells can be performed. In order to get more precise estimations for the interconnection capacitances, extracted values of the design layout can be used. If these are unavailable, wire-load models that represent worst case scenarios for synthesis have to replace extraced capacitance values [93].

The gate-level simulations have to use a sufficient number of input stimuli in order to get meaningful estimations. This leads to a considerable simulation effort for larger designs.

A complementary approach to cell level analysis are probabilistic power estimation techniques, which use statistical properties to describe the behavior of signals. There are several approaches for a statistical description of logic signals:

- using static probabilities for the state logic zero (one)
- using the transition probability under the assumption of a memoryless logic signal
- using two different transition probabilities as a function of the current state of the signal (which can be associated to a memory of length one) which is referred to as *lag-one* signal model [281]
- using a lag-N signal model
- using the static probability together with a lag-zero, lag-one or a lag-N signal model (with increasing computational complexity)

A signal that is described by one of the statistical properties mentioned above can be used to *propagate* the statistical properties of the inputs through combinational logic yielding the statistical properties of the output(s).

Given a Boolean function $y = f(x_1, x_2, ..., x_n)$ and the static probability (for logic 1 without loss of generality) $P(X_i)$ as well as the (lag-zero) transition density $D(x_i)$ the statistical properties of y can be calculated using Shannon's decomposition of the function f which is

$$y = x_i f(x_i = 1) + \bar{x}_i f(x_i = 0) \tag{3.12}$$

The static probability of this decomposed representation yields

$$P(y) = P(x_i)P(f(x_i = 1)) + (1 - P(x_i))P(f(x_i = 0)) \tag{3.13}$$

This decomposition can be recursively evaluated until the function f is completely decomposed.

A similar approach can be made for the transition density $D(y)$: a transition of y as a response to a change of x_i occurs, if $f(x_i = 0) \neq f(x_i = 1)$. This condition which is also called the *Boolean Difference* of y w. r. t. x_i can be expressed as an exclusive-OR of the two functions:

$$\frac{dy}{dx_i} = f(x_i = 1) \oplus f(x_i = 0) = 1 \tag{3.14}$$

The probability of a transition of y due to a transition of x_i is given by the product of the static probability for which 3.14 is valid and the transition probability of $D(x_i)$. Iterative application of this formula yields

$$D(y) = \sum_{i=1}^{N} P\left(\frac{dy}{dx_i}\right) D(x_i) \tag{3.15}$$

This propagation of statistical signal properties is used e.g. by commercial tools like Synopsys' DesignCompiler [235] for internal nodes that have not been annotated with static probability and switching activity.

The purpose of low-power or low-energy design techniques is to find an architecture for a given application or a set of applications that represents the optimum concerning the "power or energy consumption". For a reasonable comparison of different architectures concerning power or energy consumption, several metrics can be used:

- plain power in *mW* to describe e.g. the average or peak power of an architecture
- plain energy in *mWs or mJ* which describes the (average) energy consumption of a given architecture to perform a given application

An additional metric which is often used for (circuit-level) VLSI design is the *power delay product.* This metric can be viewed as energy per computation, which expresses the energy-efficiency of an implementation for a given task. The result of the considered computation is available after the delay, which is part of this metric, and the total energy of this computation is the average power times the delay. Another interpretation of this metric makes sense, if voltage scaling or other techniques are used that have an impact on power and delay: reduced voltage reduces the power quadratically whereas the delay is typically (nonlinearly, refer to Equation 3.11) increased. This metric compensates the effect of the power decrease with the effect of the delay increase to get an equal weight both for power and delay. A non-equal weight is included in the *energy delay product* [114], which is equal to a 'power-delay-delay' product; this metric is useful for applications that favor processing speed over energy consumption.

Various other metrics have been proposed and are used for different purposes, specifically for μ-processors, where the application is not fixed:

- Mega Instructions per mW (*MIPS/mW*) can be used to compare different μ-processor implementations that have the same or a very similar instruction set with a typical benchmark application
- other metrics use *operations* instead of *instructions* e.g. [37] uses power per throughput (in mW per operations per second) and energy per throughput for fixed throughput and maximum throughput operation as well as a metric, which normalizes total energy consumption to the maximum throughput scenario

The choice of the power metric strongly depends on the optimization goal. For many portable applications with fixed processing rates (which corresponds to fixed throughput) constrained by the application (e.g. speech, mobile reception/transmission etc.) the metric simply has to maximize the battery lifetime. In such a case a metric *energy per computational task* or *energy per typical set of operations* is suitable, which has to be interpreted as the above-mentioned *energy per operation* for a given benchmark. For many non-battery operated appliances, the metric *average power* is often sufficient in order to keep package and cooling costs under control. However, a reduction of the average consumed energy leads to a reduction of the average power and vice versa. Therefore, these distinct metrics can be indirectly optimized simultaneously by just considering the average energy per computational task. In the following discussions and also in Chapter 7 the metric *average energy per computational task* is used as optimization goal and the terms *power* and *energy optimization* are used as synonyms.

3.3 Techniques to Reduce the Energy Consumption

The focus of this section is to characterize techniques to optimize the energy per computational task starting with a high level behavioral implementation of the algorithm and ending with a synthesized netlist of standard cells. Figure 3.2 depicts the different levels of abstraction: the impact of power saving techniques decreases with increasing level of implementation detail, whereas the accuracy of power estimation increases. This is an issue that makes it difficult to predict the effect of e.g. algorithmic changes on the power consumption.

It has to be mentioned, that the following classification into system/architecure level, logic and physical level is not orthogonal for all low-power techniques: there are techniques which affect more than one level in this hierarchy. This fact emphasizes the importance of joint power optimization on all levels of abstraction.

Furthermore, the techniques described in this section are typically restricted to semi-custom design flows – special circuit techniques enabled by full-custom design are not covered. An exception is the important technique of voltage scaling, which is not commonly used for

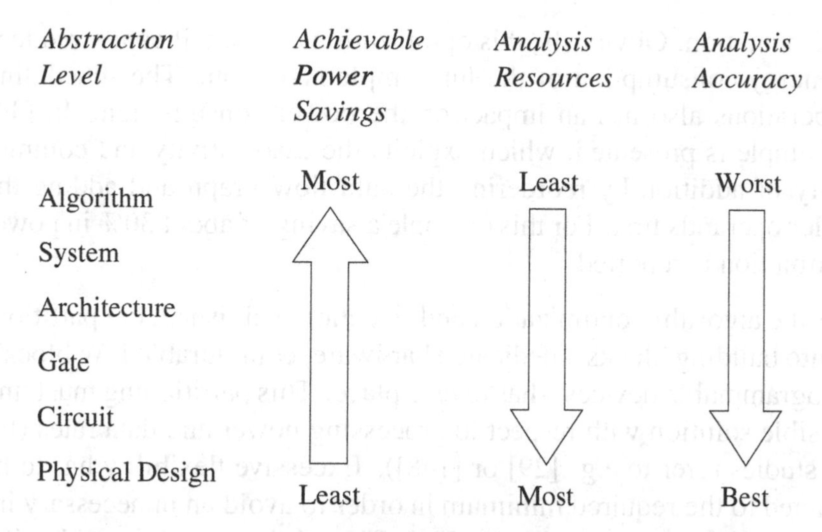

Figure 3.2: Level of Abstraction vs. Possible Savings (Irwin [124])

current semi-custom design flows. However, this technique might become important in the future and is therefore included in the following discussion.

3.3.1 System and Architecture Level

Given a certain application, the first choice in the design flow is the selection of an optimum algorithm with respect to the *cost function* of the design. The term *cost* depends on the application and typically includes the number of operations (additions, multiplications, logic operations etc.), the number of memory accesses as well as the memory size. The decisions on this level of abstraction typically have a large impact on the design efficiency. Unfortunately, the estimations on this level of abstractions tend to be significantly inaccurate unless a complete implementation in a high level language is available and a complete floating point to fixed point conversion has been done. After this design step (which has to be performed for all considered algorithms) more precise values for the complexity of memories, arithmetic, and logical operators can be estimated. The traditional purpose of this algorithm optimization is the **reduction of operations**, **memory accesses** and **memory size** in order to reduce the area and to increase the computational throughput of an implementation. Obviously, this optimization also significantly reduces the energy consumption of the final implementation. The scheduling of operations also has an impact on the power consumption. In [44] an example is presented, which exploits the associativity and commutativity of addition by reordering the data flow graph and adding the smaller operands first. For this example a saving of about 30% in power consumption is reported.

After the algorithm optimization and selection is finished, the partitioning into building blocks – dedicated hardware, configurable HW blocks, or programmable devices – has to take place. This partitioning must find a feasible solution with respect to processing power and data rates (for case studies refer to e.g. [29] or [148]). Excessive flexibility has to be restricted to the required minimum in order to avoid an unnecessary increase in power consumption [1] [92]. Thus, it is important to identify, whether the amount of required flexibility of a building block can be satisfied with (coarse-grain configurable) dedicated hardware. An important parameter for this partitioning is obviously the computational performance of the considered task. Moreover, parameters like area efficiency for low data-rate tasks and flexibility requirements for error-prone and quickly changing control tasks have to be taken into account [91].

It is possible for some algorithms to use adaptive implementations, where the **number of operations** that are needed for this task can be scaled to reduce energy. This typically also affects the algorithmic performance (e.g. bit error rate, mean square error etc.). However, if the application permits a certain algorithmic degradation under some circumstances, it might be advantageous to detect this condition and scale the algorithm accordingly. Theoretically, any iterative algorithm is a candidate for this saving technique provided that the overhead of estimating the scaling criterion is small compared to the expected savings. One application for such a technique are the adaptive filters in [172] [144], where the signal-to-noise ratio is estimated and used to adapt the filter length of the FIR filter. Another example monitors and controls the progress of iterative matrix diagonalization by low overhead techniques [147].

An often-used technique for lowering the power consumption is **voltage scaling** typically in combination with parallelization of hardware units. This techniques has already been raised in Section 3.2.2. If the initial algorithm is easily parallelizable or pipelinable, this technique is straightforward. However, many algorithms are inherently sequential due to data dependencies, which makes parallelization more difficult if not impossible. In such a case it might be worth changing the algorithm to an approximation that exhibits higher parallelism like in the well-known case of Turbo decoders [173]. On the other hand, the sequential description of an algorithm can be modified without changing the output behavior of the algorithm in order to exploit more parallel operations e.g. by loop unrolling [45].

Another approach for tasks with low or non-existent throughput constraints is the reduction of the supply voltage without a change in implementation tolerating a certain degradation of computational performance. This technique has been used in [206] to scale the voltage dynamically for a microprocessor system by using a power-conscious operating system. However, for many DSP applications with fixed throughput requirements this approach is infeasible.

Memory accesses are expensive in terms of energy consumption, because heavily loaded internal bit- and word-lines have to be switched. The average energy consumption of a memory (read or write) access increases with the memory size. To make matters worse, external memory accesses require switching even higher pad and external capacitances. Therefore, algorithmic transformations in order to reduce the number of memory accesses and/or to reduce the memory size are also effective power saving techniques on the system level (cf. [85], [181] and [42] for examples). Accesses to large memories should be reduced by using an appropriate memory hierarchy: starting with registers as the lowest level of hierarchy, this hierarchy ends with large on-chip or external memory banks. Accesses to registers are obviously much less power consuming (and typically also much faster) than accesses to larger memory blocks. Favoring local over global communication in this example, enables to decrease the power consumption.

A well-known technique for many μ-processors is applicable for any kind of low-power hardware: **power management**, which shuts down inactive parts of the chip. This can be done on different levels e.g. by gating the clock for unused parts of the circuit (which is automated by Logic Synthesis Tools like the DesignCompiler [235]) or even by entirely shutting down the clock generation unit in the phase-locked loop like in [73]. Power management for complete modules on the chip requires either software support e.g. provided by a power-conscious operating system or a dedicated hardware controller.

For programmable architectures, the overhead in terms of instruction fetching, instruction decoding, data routing etc. can be reduced by increasing the number of useful operations per time unit without increasing the overhead energy. In [37] it is stated that VLIW architectures are the best candidate for this optimization because they exploit instruction level parallelism (ILP). However, real VLIW implementations tend to increase the overhead energy significantly due to larger instruction memories and decoders. This disadvantage can be partially reduced by instruction compression techniques to reduce the instruction memory width by avoiding the explicit coding op no-operations opcodes in the instruction. Furthermore, more elaborate compression schemes reduce the redundancy of programs by exploiting statistical properties. Examples for instruction compression techniques are the simple fetch scheme of the commercially available TMS320C62xx [246] and also more elaborate techniques used in academia like in [163]. The simple scheme of the TMS320C62xx, however, results in several power consuming decoding stages, which are needed to decode and route instructions to functional units. On the other hand, more elaborate compression schemes result in typically significant hardware effort and energy to decompress the code due to large look-up tables. A completely orthogonal technique has been used in the case study of Chapter 7.1, where application-specific instructions have been implemented to increase the number of parallel operations per instruction *without* a significant impact on the overhead energy.

3.3.2 Register Transfer and Logic Level

Low-power techniques on the register transfer (RTL) and on the logic level can be subdivided into techniques for lowering the capacitance and the switched voltage as well as into techniques to reduce the toggle rate of nodes with a high relative capacitance. Furthermore, toggle activity for un-useful calculations should be reduced to a minimum. Reduction of the switched voltage is beyond the scope of this thesis, because it requires special circuit techniques that are (so far) not applicable to semi-custom design flows.

Lowering the capacitance can be achieved by reducing or avoiding global communication as far as possible because global communication implicitly requires switching long interconnections with high capacitance. However, for heterogeneous systems using different layout blocks on a single chip it is often unavoidable to use long interconnections. In such a case the interconnection network has to be reduced to a minimum and the topology should favor point-to-point or nearest-neighbor connections [1]. For the same reason, external communication should be reduced to a minimum e.g. by using internal cache memories. Much effort of lowering capacitances is used by the synthesis tool, which implicitly reduces the switched capacitance by logic optimization targeting minimum area and in many cases also if targeting maximum speed. Advanced techniques to explicitly reduce power by optimum technology mapping are reported in [248].

If the capacitances can not be further reduced, the orthogonal approach is to reduce the switching activity of interconnections with high capacitances. Various approaches have been described in the technical literature. The most popular technique – clock gating – can be classified as a so-called **guarding technique**. Clock gating means to shut down the clocking for a certain group of registers under a certain *guard condition*. An obvious example for this technique is to shut down the clocking of pipelined functional units in a microprocessor e.g. in [223]. Clock gating techniques on a more fine-granular level are possible like in Figure 3.3 [45], where the input of a comparator is guarded against the trivial condition that the MSBs are different to avoid the evaluation of the full input word lengths in this case. This special guarding logic together with the MSB comparator is also called *precomputation logic*, because the result can be quickly precomputed using a subset of the circuit inputs. Furthermore, guarding techniques to avoid propagation of data values into functional units that are connected to a common bus [91] [251] are extremely efficient, because they can typically be implemented with minor overhead in area and design effort. So-called *extended guarding techniques* [251] comprise conventional guarding logic as well as additional logic that can be viewed as precomputation logic. Guarding techniques involve several issues with common standard cell

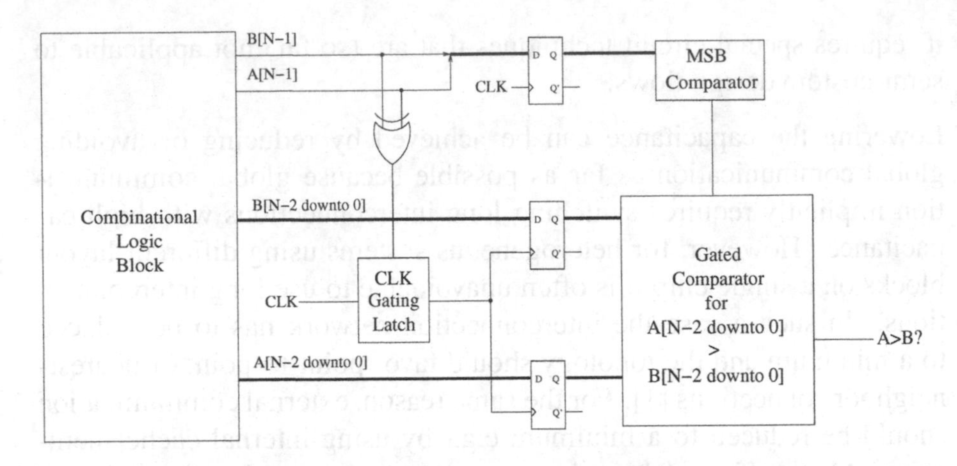

Figure 3.3: Gated Comparator

design flows: Firstly, the use of latches is typically prohibited due to testability issues. Secondly, the efficiency of guarding techniques typically depends on the timing constraint that the guarding condition is stable *before* the signals that have to be guarded are stable. The latter relative timing constraint makes timing verification and physical design significantly more complicated.

Pipelining of combinational logic has several effects: firstly, the critical path of the (synchronous) implementation is shortened, which enables savings due to voltage scaling or due to slower implementations of arithmetic operations with a higher energy-efficiency (a comparison of arithmetic implementations is given in [40]). Secondly, glitches (also called spurious transitions) within the combinational logic due to unbalanced signal propagation are reduced, which also results in lower energy. Unfortunately, pipeline registers in semi-custom technology are typically extremely costly in terms of area (with an implicit increase of capacitances due to higher distances on the chip) and, to make matters worse, increase the clock power (in the clock tree as well as in the register circuits). This negative effect on power consumption has to be compensated with clock gating, wherever this is possible. **Retiming** can also be used to reduce the area penalty of pipelining to a certain extent as well as to reduce the switching activity of logic nodes [197].

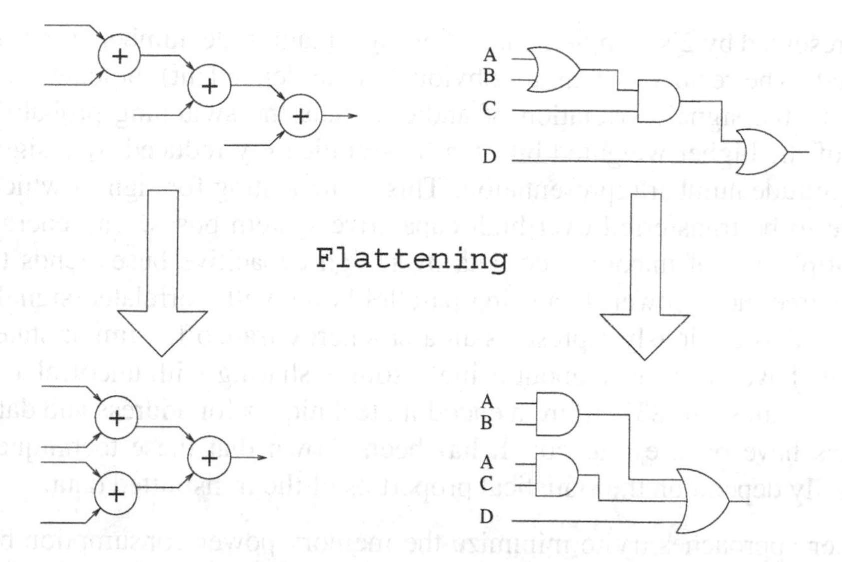

Figure 3.4: Flattening of Operators and Logic

Further reduction of switching activity on highly capacitive nodes due to glitching can be achieved by **reorganization of logic gates and operators** [45] [142] like in the examples of Figure 3.4. Reorganization of operators has to be typically performed manually but reorganization of logic cells and also reordering of equivalent inputs [281] can be automatically performed by commercial synthesis tools [64]. The optimization tasks of the logic synthesis tool can be subdivided into combinational optimizations like

- don't care optimization [214]
- path balancing
- factorization

as well as sequential optimization like

- state encoding
- retiming

The data representation itself also has an impact on the switching activity: in [45] the transition probability for each bit of 16 bit audio data represented by 2's complement and by sign-magnitude numbers is compared. The results (which are obviously data dependent) indicate, that due to the signal correlation of audio signals the switching probability of the higher weighted bits can be significantly reduced by a sign-magnitude number representation. This is interesting for signals which have to be transferred over high capacitive system buses. In general, multiplexing of uncorrelated data over high capacitive buses tends to consume more power than using parallel buses with correlated signals [45]. This obviously represents an area-energy tradeoff. Similar statements have been made about using resource sharing with uncorrelated data streams. In [83] different encoding techniques for address and data buses have been evaluated. It has been shown that these techniques heavily depend on the statistical properties of the transmitted data.

Other approaches try to minimize the memory power consumption by using runtime compression techniques in combination with intelligent memories [185].

In Chapter 7.1.3 the effect of minimizing the internal power of an instruction memory by reducing the number of discharging events in the instruction ROM is described. This minimization has been performed automatically using instruction-frequency-driven maximum weight encoding[1]. The tools that have been developed for this optimization (refer to Subsection 6.3.1 for details) can also reduce the switching activity of the (external) instruction bus, if this is desired. A more limited approach is described in [273] where the don't care bits in a microprogrammed control unit are optimally assigned using trace driven activity evaluations.

3.3.3 Physical Level

On this level of abstraction the number of manually guided optimizations is quite limited because the semi-custom design flow uses automatic place and route tools in order to transform the netlist of standard cells into a chip layout. The place and route tools automatically minimize the wire length (and wire capacitances) according to the time constraints. However, this does not necessarily represent the optimum con-

[1]This technique has been developed and published in [129] [90].

cerning power consumption, because the switching activity is typically not taken into account. Automatic gate sizing using in-place optimization (IPO) with area and I/O standardized buffer cells can be used to optimize the transition times of logic as well as clock nodes after an initial place and route pass has been performed.

There are some design tasks which can nevertheless be exploited to save power on this level of abstraction: partitioning and back-annotating of layout information to the synthesis tool.

Partitioning and floorplanning for low-power can be done taking into account the interconnections between the layout blocks, which are typically defined earlier in the design process (normally during architecture design). The length of interconnections with high switching frequency should obviously be minimized which corresponds to a minimization of the distance between the associated layout blocks. The I/O ports of physical blocks may have to be manually defined in order to achieve optimum interconnections.

Back-annotating of layout capacitances together with the switching activity information from gate level simulation to the synthesis tool can enable efficient **reoptimization of logic** for low-power. This technique has already been described in the previous subsection.

3.4 Concluding Remarks

This chapter summarizes the metrics for efficient hardware implementation and efficient hardware design. Furthermore, the sources of energy consumption of state-of-the-art CMOS technology are described. Moreover, a concise summary of low-power design principles as well as specific design techniques in order to lower the power or energy consumption is given. Many of these techniques are used in the ASIP design flow described in Chapter 5.

Chapter 4

Application-Specific Processor Architectures

The ASIP design space classification presented in this chapter identifies the degrees of freedom in the ASIP design process. This discussion neglects low-level hardware implementation details, which have been treated in the previous chapter; it rather uses the abstraction of word level hardware operators like addition, muliplication, etc. The design decisions on this level of abstraction significantly affect the resulting architectural efficiency as well as the overall design efficiency. As a consequence, this chapter is a prerequisite for the ASIP design flow presented in Chapter 5. Furthermore, this classification enables to decide, if a certain architecture can be supported by an available high level language compiler design environment or a retargetable HLL compiler to enable high level language support. Finally, this chapter treats the important relation between high level design decisions and critical low level implementation characteristics. This relation has to be well understood in order to obtain optimum design results.

This chapter starts by defining important terms in the context of ASIP design and embedded signal processing architectures. Afterwards, several important fields of ASIP applications are discussed together with references to ASIP case studies. Finally, the design space of ASIPs is defined and the impact of high level design decisions on performance, energy and area consumption is described.

4.1 Definitions of ASIP Related Terms

The technical literature uses the acronym *ASIP* to describe two different kinds of integrated digital circuits:

- Application-Specific Integrated Processor: This term represents any kind of application-specific digital integrated circuit used for data processing and does not imply any kind of instruction set oriented or programmable data processing [209].

- Application-Specific Instruction Set Processor or Application-Specific Instruction Processor[1]: This term represents a programmable application-specific processor using the concept of an instruction set architecture for data processing.

In this thesis, the term ASIP exclusively refers to an instruction set oriented processor with application-specific optimizations including optional tightly-coupled hardware accelerators.

Figure 4.1 depicts typical classes of hardware implementation paradigms, which are bounded by pure ASICs on the left side and by general purpose processors on the right side. ASIPs can be viewed as a tradeoff between non-programmable application-specific integrated circuits (ASICs) and domain specific signal processors (DSSPs). ASIPs are optimized for just *one* signal processing application. DSSPs are instruction set oriented processors targeting a complete domain of signal processing applications (e.g. a network processor optimized for a class of different network processing tasks). Conventional off-the-shelf DSPs are less application-specific and target an even broader range of signal processing applications.

The term **instruction set architecture (ISA)** defines the part of a processor that is visible to the programmer or compiler writer [107].

The term **processor architecture (PA)** extends the scope of an ISA by adding implementation characteristics that are hidden to the software: this chapter discusses PAs on the abstraction level of word parallel hardware operators. A PA contains a description of the processor resources (functional units and storage elements), of the interconnections between those resources and of the encoding/behavior of the supported instructions. The instruction behavior determines the processor's state transitions and the resource utilization of functional units.

[1] In the technical literature the term *Application-Specific Programmable Processor (ASPP)* is a synonym for an ASIP in this sense, cf. [143]

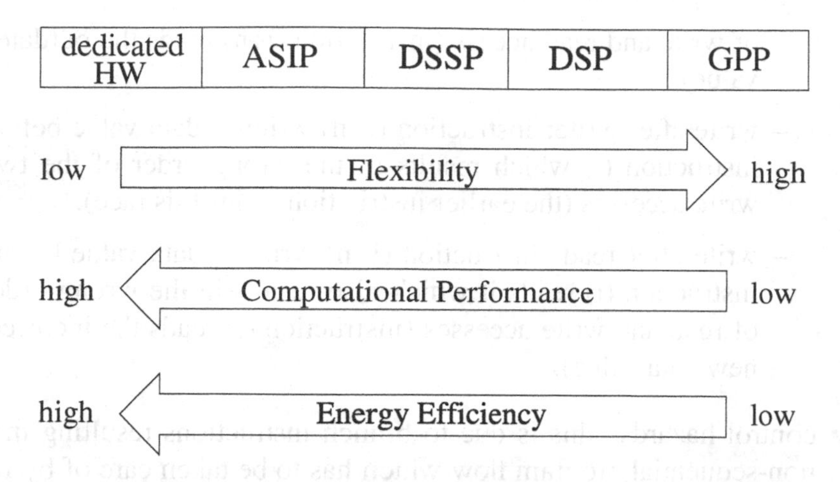

Figure 4.1: ASIPs in the context of Processor HW implementation classes

Each PA contains a **data path**, which comprises the functional units and storage elements for data processing. The architecture's remaining parts constitute the **control units**, which control the data path like the instruction decoder or the interrupt control logic. This distinction is not applicable to units that exhibit data- as well as instruction-dependent behavior, such as the branch prediction unit.

A **pipelined processor** uses pipeline registers to subdivide computational tasks into a sequence of overlapping, subsequently performed subtasks. Each subtask is executed using the combinational resources of a a so-called **pipeline stage**. Dependencies between subsequent instructions and resource conflicts result in **pipeline hazards**. Pipeline hazards are due to the dependence between instructions that are close enough so that the overlapping execution in the pipeline leads to a different access sequence of resources or data than in the case of non-overlapping execution. Hennessy and Patterson [107] classify these pipeline hazards as follows:

- data hazards: data dependencies between two instructions which would result in an incorrect behavior, if not properly resolved
  - read after write: instruction (i+n) tries to read a data value before instruction (i) writes it, which results in the wrong order

of write and read access (instruction (i+n) reads the outdated value).

  - write after write: instruction (i+n) writes a data value before instruction (i) which results in the wrong order of the two write accesses (the earlier instructions wins this race).
  - write after read: instruction (i+n) writes a data value before instruction (i) reads it which also results in the wrong order of read and write accesses (instruction (i) reads the incorrect new data value).

- control hazards: this is due to branch instructions resulting in a non-sequential program flow which has to be taken care of by inserting pipeline bubbles (insertion of "no operation" into a stage) or by using the so-called "branch delay" slot(s). Branch delay slots are instructions after the actual branch instruction that are always executed, regardless if the branch was taken or not.
- structural hazards: are due to resource conflicts of functional units or of the memory and typically result in pipeline bubbles as well.

4.2 ASIP Applications

Typical applications of ASIPs can be subdivided into the classical domains, where traditional μ-controllers and programmable digital signal processors (DSPs) in combination with dedicated hardware are used. In the last few years, a trend towards multi-threaded network processor (NPs) architectures optimized for network routing and switching applications can be observed.

Application classes for ASIPs can be subdivided into

- control-dominated systems which react to (typically non-periodical) external events often with real-time constraints on the response time
- data-dominated systems where complex transformation of data are performed using

  - cyclostationary processing of data streams (operation sequence is largely defined at compile time)
  - non-cyclostationary processing (operation sequence is strongly data dependent)
- a mixture of control- and data-dominated systems

Examples for control-dominated systems are the above-mentioned network processors whereas typical cyclostationary processing of data streams can be found in many digital processing algorithms e.g. for filtering and equalizing data or for channel decoding [48]. Non-cyclostationary data processing is typically also a part of digital signal processing systems and can be found e.g. in digital receiver structures [179] that make use of different channel acquisition and tracking algorithms.

From an ASIP centric point of view, the historical development of traditional fixed DSPs can be regarded as the continuous attempt to find the optimum fit between the feasible hardware effort and the cost of a DSP on one hand, and the demands of quickly changing, popular applications on the other hand. This slowly developing process of DSP evolution has produced ASIP-like features in general purpose DSPs like e.g.

- single cycle multiply-accumulate using the data bus and the program bus as sources for the multiplier (TMS320C2x [244])
- bit-reverse addressing mode e.g. for FFT-butterfly addressing (TMS-320C2x and C54x [245] and many others)
- subword parallelism (corresponding to a SIMD extension) using two 16 bit numbers within a 32 bit word in order to perform 2 multiply-accumulate operations in one cycle (Lucent 16000 [27] and others)
- computation of a parallelized 2-unfolded FIR or IIR using a delay register (Lode DSP [28])
- Viterbi extension for Add-Compare-Select in combination with dedicated storage for the survivor path (TMS320C54x [245])
- software pipelined Viterbi execution using two specialized Viterbi instructions (StarCore [186])

- further SIMD extensions for filtering purposes (cf. TMS320C62x, TMS320C67x [246] and TigerSharc [10])

For a limited number of algorithms e.g. FIR/IIR-filters, FFT, distance calculations or even matrix operations it is obviously possible to optimize a fixed DSP instruction set architecture prior to fabrication. However, if quickly evolving applications call for significantly different algorithms, these "optimized" DSPs might expose poor performance. In the worst case, an application might need an optimum implementation for a mixed control- and data-dominated task, which calls for a mixed implementation using features of μ-controllers together with application-specific DSP features like in [189]. In such a case, a reasonably designed ASIP that is solely optimized for the underlying application will certainly outperform available fixed DSPs and μ-controllers.

In Table 4.1 some commercial and academic ASIP case studies are listed as examples for typical ASIP applications. However, ASIP design has been common in the industry for a longer period of time in the form of in-house DSPs, which are intended for a specific application domain [201].

4.3 ASIP Design Space

The following classification of processors focuses on architectural features that are relevant for the implementation of ASIP processor architectures.

Flynn's classification [82] is the most popular and lucid processor classification based upon the number of instruction and data streams that can be simultaneously processed. The processor categories are:

- SISD (Single Instruction, Single Data), which is the the classical definition of a scalar uniprocessor.
- SIMD (Single Instruction, Multiple Data), which defines the class of vector/array processor.
- MISD (Multiple Instruction, Single Data) is often considered irrelevant in practice. Nevertheless, instruction level parallel architec-

Application	Authors, Affiliation and Reference	Design Environment
MPEG-I decoding	P. Plöger, J. Wildberg GMD [207]	CASTLE
MPEG-II enc., LMS Adaptive Filtering	S. Balakrishnan et al. Univ. of Twente [21]	SYMPHONY
UNIX "crypt"	V. Zivkovic et al. [283]	MOVE
Java Processor	Serfio Akira Ito et al. UFRGS - Brazil [125]	-
ATM cell processing	S. Virtanen et al. TUCS Finland [265]	TACO
Vector Processing	M. Gschwind IBM Research Center [101]	-
MD5 encryption SHA	P. Faraboschi et al. HP Lab. and STM Cambridge (MA) [77]	Lx platform
JPEG2000 among others	D. Chuang Improv Systems Inc. [51]	Improv Design Platform
FIR, JPEG, Viterbi, Motion Detection, DES	R. E. Gonzales Tensilica Inc. [97]	XTENSA Proc. Design Platform
RISC+DSP applications	ARC Cores Ltd. [12] [13]	ARC ARCtangent-A4 DSP

Table 4.1: ASIP Case Studies

tures (like VLIW or superscalar architectures) with a non-parallel load/store unit and a single I/O port are part of this class.

- MIMD (Multiple Instructions, Multiple Data), which covers the range of many instruction level processors and multi-processor/computer systems.

Flynn's classification is a good starting point for the following design space definition to differentiate between the non-parallel SISD architectures and the parallel SIMD and MIMD architectures. In Figure 4.2 a classification of parallel architectures (similar to [227] with minor modifications) is depicted.

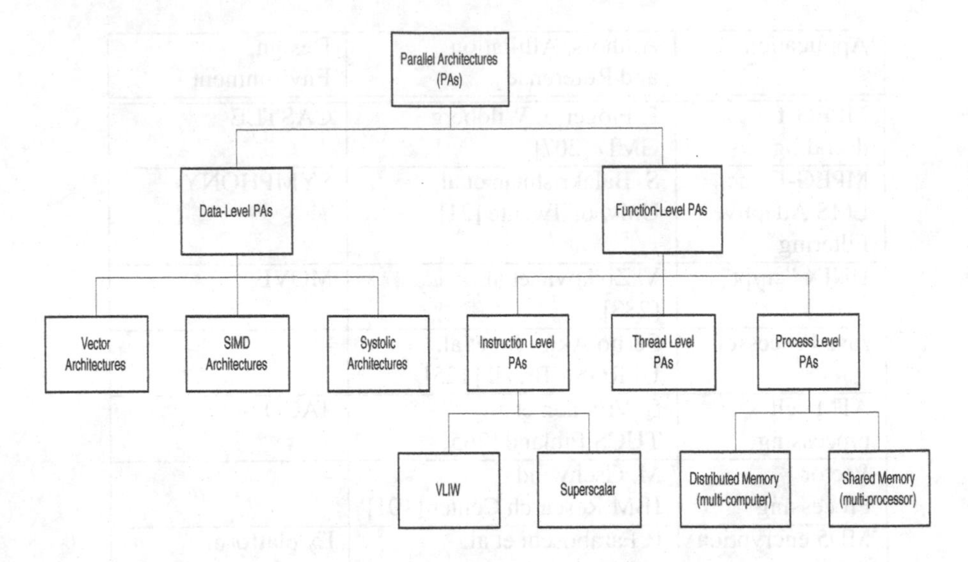

Figure 4.2: Design Space of Parallel Architectures (Sima [227])

According to this classification, typical DSP applications like FIR, vector or matrix computations obviously represent a good match with data-level parallel architectures. In fact, efficient hardware implementations of these algorithms use dedicated hardware structures, which resemble the data paths of these instruction set architecture classes (mostly with further application-specific optimization like in [48]). The SIMD principle is also used by some commercially available DSPs (e.g. the C6X DSP from Texas Instruments [246] or the Trimedia Processor from Philips [205]) by implementing SIMD instructions to support multiple parallel operations on register subwords. For high level languages the compiler has to "vectorize" the code in order to target these architectures efficiently. This vectorization is difficult for high level languages like C and C++ without explicit support of vector and matrix operations. This is one reason, why VLIW architectures, which avoid this issue, have become popular both for general purpose (e.g. for Intel's EPIC architecture [157]) and digital signal processing applications. A good overview of this topic is given in [122].

On the other hand, superscalar processors tend to have significantly more complex hardware, which is needed to exploit instruction level parallelism during program runtime. This extra hardware also needs significantly more silicon area and energy consumption, which is prohibitive for energy critical, embedded digital signal processing applications.

Multi-threaded processors are used in particular in the area of network processors. Multi-threading can be generally applied to utilize the functional units of a processor more efficiently. This concept typically is beneficial to hide memory latencies in order to increase the processor's throughput without affecting the computational latency for a single thread. For tasks with regular data access patterns in time critical tasks, however, conventional DSPs with optimized memory organization are often more suitable.

Process-level parallel architectures and systems are common for embedded systems in order to balance the workload of one processor. Typically, a combination of distributed memory with shared memory or dedicated inter-processor communication resources is used to avoid communication bottlenecks for number crunching algorithms [213].

The taxonomy in Figure 4.2 still lacks many important architectural details, which are of practical relevance for ASIPs. The following subsections classify the ASIP design space with respect to important architectural features. Each subsection describes one group of related, but orthogonal design parameters.

4.3.1 Functional Units

The functional units represent the data paths's elements of a processor. The following characteristics and parameters can be identified for the functional units:

- granularity: bit serial, word serial or word parallel operation
- word width, number of parallel words etc.
- arithmetic: fixed point, block floating point or floating point
- operation(s): e.g. integer arithmetic, complex arithmetic, boolean operations, Galois field operations, etc.

- configurability: fixed operation (e.g. signed multiplier) or configurable operation (signed and unsigned multiplier)
- single vs. multi-cycle operation

These characteristics are sufficient to span the design space for the behavior of the functional units. Further aspects like control and pipelining of these units are covered later on.

4.3.2 Storage elements

Storage elements in a processor system (including data and instruction memories) are used to temporarily store data and control information. The following list of characteristics determines the organization of storage elements for an ASIP:

- word width and number of addressable words in the storage element
- register organization: orthogonal register file, split register files or distributed registers
- location of memory: on-chip memory or external memory
- access time of memory: number of processor cycles to read/write data
- memory organization: one memory for instructions and data (von Neumann architecture [192]) or different memories for instructions and data (Harvard architecture) [55]
- memory hierarchy: flat instruction memory or hierarchical organization (using caches)
- data memory organization:
  - single or dual ported memories
  - single memory bank or several data memory banks for simultaneous access
- instruction memory parallelism: sequential read of instruction or parallel fetch of several instructions

- instruction memory type: synthesized or hard-macro ROM, boot-loadable RAM or a combination of ROM and RAM

This classification includes the well-known register-register architectures (which use a data register file) as well as the register-memory and the memory-memory architectures (typically with distributed internal data registers).

An orthogonal aspect of storage elements is how the processor accesses them. For register, which are connected to just one functional unit, this access is straightforward, because it is determined by the dedicated connection of this register. General purpose register files offer a limited number of read and write ports and are often connected to data/address buses or multiplexer structures. Data memory accesses are typically controlled by special load/store units. Depending on the data memory organization, one or several simultaneous read/write operations can be performed. Accesses to the same memory bank have to be restricted by the load/store unit to just one access (two accesses) per cycle for single (dual) port memories.

4.3.3 Pipelining

The concept of pipelining in a processor can be applied to single combinational functional units or to subdivide groups of functional units into different stages for instruction execution. The concept of pipelining is not orthogonal to the organization of storage elements in the previous subsection, because it introduces additional storage elements to the architecture. The purpose of additional pipeline register is not primarily to store data rather than to increase the computational performance (sometimes also to increase the energy-efficiency like in Section 7.1).

Pipelining of single combinational functional units increases the maximum clock frequency of this unit and, thus, increases the possible computational throughput. This is especially useful, if the same computation has to be performed for a series of input data. Pipelining is also a technique to utilize functional units more efficiently, because a computation is partitioned into subcomputations that are executed in parallel for a series of input data in analogy to the concept of an industrial assembly line.

Pipelining can also be used on a coarser grain of abstraction to separate different groups of functional or control units from each other. A typical pipeline organization of a RISC processor uses the pipeline stages *instruction fetch*, *instruction decode*, *read operand*, *execute* and *write-back operand* (cf. Figure 4.3). Pipelining enables higher operating frequencies. On the other hand, data and resource dependencies of different stages lead to pipeline hazards, which effectively reduce the utilization of the pipeline stages. For a more detailed discussion of pipeline hazards refer to Section 4.1 and [107].

Instruction Number \ Cycle Number	1	2	3	4	5	6	7	8
1	IF	ID	RD	EX	WB			
2		IF	ID	RD	EX	WB		
3			IF	ID	RD	EX	WB	
4				IF	ID	RD	EX	WB
5					IF	ID	RD	EX
6						IF	ID	RD
7							IF	ID
8								IF

Figure 4.3: Example RISC Processor Pipeline

The total processing time T_{pipe} to process n instructions with a linear pipeline of s stages is

$$T_{pipe} = (n + s - 1)T_{clk} \tag{4.1}$$

for a clock period of T_{clk}. In the following ideal consideration, pipelining overhead due to setup times of real flip-flops is neglected. In the limiting case of identical critical timing paths T_{clk} of each pipeline stage the equivalent unpipelined architecture needs

$$T_{unpipe} = nsT_{clk} \tag{4.2}$$

for the same computation.

As a result the speedup factor of pipelining is

$$S_p = \frac{T_{unpipe}}{T_{pipe}} = \frac{ns}{(n+s-1)} \tag{4.3}$$

If the additional area for the pipeline registers is neglected, pipelining leads to an increased architectural efficiency (cf. Section 3.1.1):

$$\eta_{arch,pipe} = s\,\eta_{arch,no\_pipe} \tag{4.4}$$

4.3.4 Interconnection Structure

The interconnection between functional units and storage elements determines the flexibility of a data path, which is the most important distinguishing feature between more dedicated and general purpose data paths.

Basically, there are two options for the interconnection between two nodes: unidirectional or bidirectional interconnection. Unidirectional interconnection is implemented using a simple wire between the output of the consuming and the input of producing node. Bidirectional interconnection are more complicated, because the designer has to make sure that the required bidirectional drivers are not simultaneously driving the bus with different logic values, which would result in short circuits. This problem can be avoided by using separate input and output ports for each node together with separate unidirectional interconnections between these ports. However, this approach needs more silicon area, which would be a significant drawback for system buses.

There is a large variety of possible different interconnection networks e.g. using binary trees, stars, meshes or systolic arrays [120]. However, all these interconnection networks can be constructed using two fundamental topologies:

- one output producing information for one or several inputs
- one or several outputs producing information for one input

In Figure 4.4 the hardware implementation for these two options is depicted. It is obvious that the left implementation does not require any additional combinational hardware, whereas the right implementation needs a multiplexer or tristate output drivers (with additional control units). The overhead due to interconnections (especially in the case of non-tristate buses) can be significant for a highly configurable target architecture due to excessive relative interconnection silicon area and delays of deep sub micron technologies with respect to combinational logic [19]. For that reason, the interconnection network should be carefully dimensioned preferring local over costly global communication and minimizing the interconnection flexibility as far as possible for the target application.

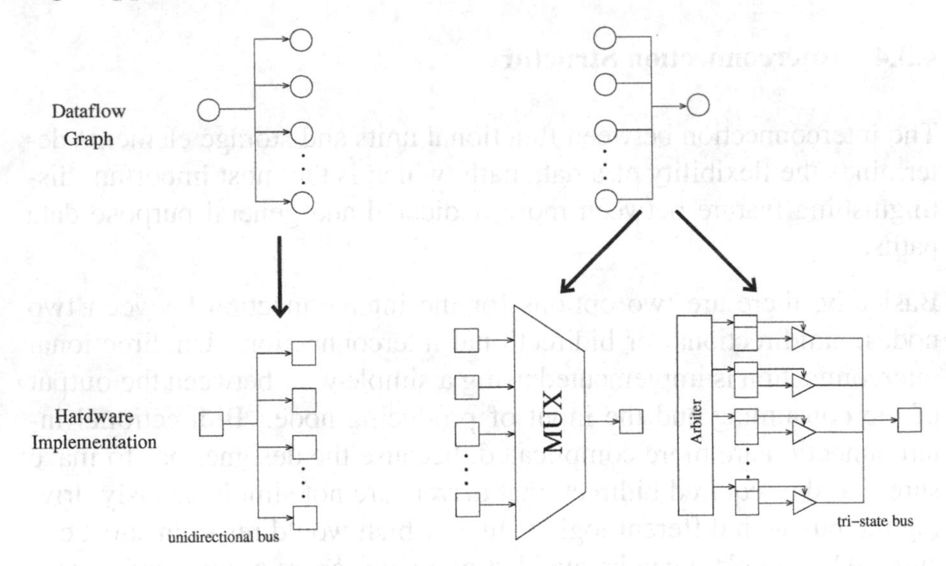

Figure 4.4: Basic Network Topologies

4.3.5 Control Mechanisms

There are two different mechanisms to control the data path of an instruction set oriented processor

- time-stationary coding: the instruction controls exactly one state transition of the complete data path
- data-stationary coding: the instruction travels together with the associated data in the pipeline and controls the sequence of operation(s) performed on these data in each pipeline stage

As reported in [98], many ASIPs use time-stationary coding, because the programming and verification of these architectures is facilitated. However, for more deeply pipelined architectures pure time-stationary coding is inefficient due to a large number of redundant configuration bits needed for the instruction. For these architectures data-stationary coding is obviously more suitable (cf. [91] for an example).

The design of pipelined data paths using data-stationary coding requires the following design decisions:

- open pipeline: The pipeline is fully visible to the programmer and the programmer has to take care in order to avoid structural and data hazards (which both would lead to incorrect program behavior). The same is valid for control hazards: the programmer has to fill the delay slot(s) with valid instructions after each control instruction.

- interlocked pipeline: The pipeline is not visible to the programmer, because the hardware takes care of structural, data and control hazards by using
  - pipeline interlocking (stalling of previous pipeline stages) in order to resolve these dependencies
  - forwarding and register/memory bypassing to avoid stall cycles by smarter data routing

For the processing of program loops, special hardware support for zero-overhead loop processing can be implemented. This hardware replaces the instructions at the end of the loop (increment/decrement of loop counter, compare with end value, branch on this condition) and avoids the associated branch penalty. This kind of hardware loop support has been implemented in many DSPs and ASIPs e.g. [186] [97] and [89].

In the last few years, conditional instruction execution has become popular for deeply pipelined processor architectures. *Conditional or predicated execution* means that the execution depends on a special condition or predicate register. This condition/predicate bit can be set by e.g. a comparison instruction to implement a HLL statement like "if (...) then ... else ..." without using conditional branches. Consequently, control hazards have been avoided using this approach.

Residual control of functional units is sometimes applied to configure e.g. the saturation/overflow mode of an ALU. This mechanism uses a dedicated control register and is especially beneficial, if changes of this residual configuration (which can be modified by a processor instruction) are rare.

Distributed or centralized control can be used for processors with distributed functional units and split registers or for multi-processor systems. As typical ASIPs mostly use simple structures with local functional units, a more thorough investigation of distributed architectures clearly is beyond the scope of this thesis.

4.3.6 Storage Access

The access methods for the storage elements of a processor can be subdivided into **register access** and **memory access** .

Register access for dedicated registers that are connected to just one functional unit are simply controlled by the instruction type (e.g. ALU- or multiply-instruction). Access to a data register file with multiple internal registers and with multiple read/write ports typically has to be controlled by special operand fields ("register" fields) of a processor instruction. The input data for a register write operation typically are produced by either a functional unit, by the memory or are extracted from an immediate field of the instruction.

Memory accesses often use more elaborate addressing schemes:

- direct or absolute addressing: the address is directly extracted from the instruction "direct address" field

- indirect or register deferred addressing: the address is taken from a (data or address) register
- indirect with displacement: same as indirect but with an additional displacement extracted from the instruction
- indexed addressing: the address is calculated using two (data or address) registers (often the effective address is just the sum of the two registers values)

Furthermore additional so called "post-"operations are often associated with the above-mentioned addressing modes. The most simple post-operation is post-increment/decrement, which is used to increment/decrement the associated address register by a constant in order to make it point to the next data address in memory. More sophisticated post-operations include the popular addition with reversed-carry chain propagation, which is used for FFTs.

4.3.7 Instruction Coding and Instruction Fetch Mechanisms

Instruction coding determines two important aspects of an ASIP implementation: the program memory size[1] and the implementation flexibility. A decrease of the instruction width obviously reduces the instruction memory width, but it also reduces the flexibility of the encoded instructions. For instance, a RISC instruction format with three register operand fields enables operations like $(R3 = R1 + R2)_{instr1}$ using just one instruction. A two operand instruction format has to use two instructions for the same operation $(R3 = R2)_{instr1}; (R3 = R3 + R1)_{instr2}$. Even for this simple example, the effect on overall instruction memory size depends on the application program. This fact is exploited by application-specific processors, where the processor designer can optimize the instruction encoding within the following two bounds (which represent extremes w.r.t. instruction width and flexibility):

- micro-coded instructions, which offer the highest possible flexibility and need the widest instruction memory (an elaborate instruction decoder is unnecessary in this case)

[1] The size of the program memory also has a considerable impact on energy consumption

- application-specific compressed instruction encoding obtained by enumeration and binary coding of all different instructions in a given program (this heavily restricts the flexibility of the supported instruction to the set of instructions, for which the encoding has been performed but yields the minimum possible instruction width)

For many practically relevant cases, an instruction coding that encodes the available operations using a fixed length instruction field is used. Furthermore, the operation's operands like register or memory operands typically have to be programmable, thus, requiring associated operand fields in the instruction. For less orthogonal instruction set architectures these operand fields can be partially omitted using partially hard coded operands for one or more instructions. This reduces the memory footprint, which can be exploited by ASIPs [91], provided that the decrease in flexibility of the ISA is acceptable.

Typically, it is a challenging task for a given application to find the optimum instruction coding that represents a feasible tradeoff between flexibility and code size.

The instruction coding also has an impact on the memory organization and on the instruction fetch stage. Instruction fetch mechanisms describe the way instructions are routed from the instruction memory to the instruction decoder. For scalar architectures with a single instruction fetch per cycle this mechanism is trivial. However, for VLIW or superscalar architectures with parallel instruction fetch, an efficient and more complex fetch mechanism is essential to keep the parallel data path busy.

Basically, there are two popular, commercially used coding schemes for VLIW processors, which impact the instruction fetch stage:

- uncompressed VLIW encoding
- various compressed encoding schemes

Figure 4.5 depicts the principle of these two different schemes. The uncompressed VLIW encoding typically uses one bit field that controls the operation of each functional unit of the data path. This results in a significant waste of instructions bits for the case of non-parallelizable instruction execution, where unused bit fields have to be explicitely filled with horizontal "no operation" patterns [76]. The example for a compressed VLIW encoding in Figure 4.5 is similar to the scheme in [153] or [246] where the "P"-bit in each instruction is used to indicate that the following instruction can be issued in parallel. The disadvantage of compressed VLIW schemes is the additional hardware effort to decompress the instructions, to allocate the associated processor resources (if a specific resource is not defined by the instruction e.g. in the case of identical, replicated functional units) and to dispatch the instructions to the desired location. This decompression step can conceptually be seen as a mapping of the compressed instruction stream to a normal uncompressed VLIW representation as depicted in Figure 4.5.

More elaborate compression schemes use compile time compression, which reduce the code redundancy by using statistical methods [164] [278] [139]. These schemes require runtime decompression by hard- or software resulting in a potential performance degradation. Furthermore, the effort for architecture design and verification might increase significantly, because runtime decompression introduces several issues e.g. more complicated branch processing, which results in an unorthogonal architecture.

For embedded applications like the DVB-T receiver of [88] the instruction memory resides on-chip as a ROM. For field reprogrammable applications, however, the instruction memory is either implemented as an on-chip RAM or external memory is used. One constraint of the coding width in case of external memories is the bit width, which is often restricted to a multiple of 8 or 16 bits for off-the-shelf external memory elements.

4.3.8 Interface Mechanisms

Input and output (I/O) mechanisms for data and control information both affect the ASIP hardware as well as the ASIP software. Communication can be performed between the ASIP and other on- or off-chip devices like processors, dedicated hardware or analog components (like e.g. AD/DA converters). The following taxonomy describes the inter-

Uncompressed VLIW Format

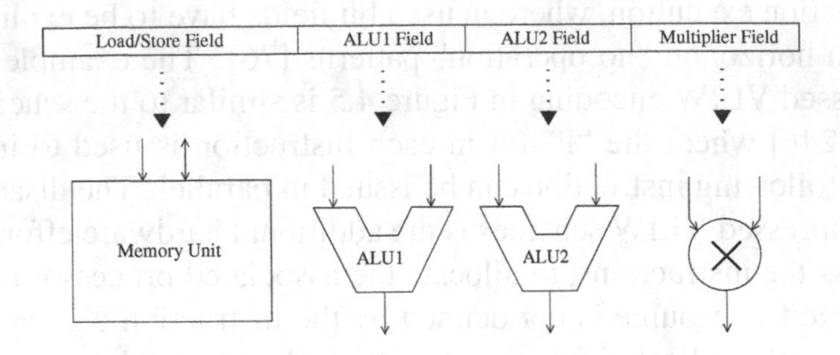

Example for Compressed VLIW Format

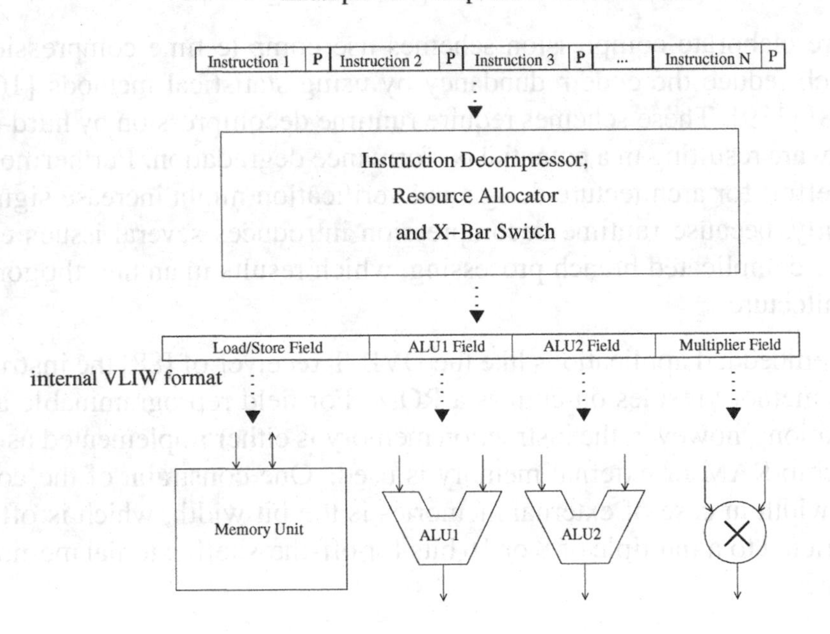

Figure 4.5: VLIW instruction formats

face mechanisms from the ASIP perspective: ASIP-external implementation characteristics of interfaces which use e.g. dedicated connections or shared system buses are not covered.

Depending on characteristics like the data rate and the number of transferred data or control samples per iteration of an algorithm (in the case of cyclostationary data processing) different I/O mechanisms can be used:

- Memory-based I/O: Data is exchanged with a shared memory. This concept is typically suitable for a larger amount of data words per iteration, which enables high data rates due to a low overhead per sample.

- Register-based I/O: either dedicated (ASIP internally read-/writeable) registers or memory-mapped registers that can be accessed similar to ordinary memory storage locations are used for data transfers. This concept is typically suitable for a smaller amount of data words per iteration, because of the large silicon area consumption of semi-custom registers. The data rate in this case is typically smaller than for the shared memory approach due to a considerable overhead per sample, which is needed to synchronize the data.

- Dedicated control channels, which typically affect the program flow for synchronization purposes (using e.g. synchronous reset signals, interrupt vectors, start-stop and/or resume-suspend signals for certain tasks including above-mentioned data transfers).

These I/O mechanisms have to be supported by appropriate hardware: e.g. in the case of shared memory, a memory arbiter has to be implemented, whereas in the case of dedicated control channels, some sort of direct memory access (DMA) controller functionality is needed like in [30]. In the case of communication based on dedicated registers, special instructions have to be implemented to access them. For memory mapped registers either a reserved I/O address space has to be used in combination with conventional load/store instructions or, alternatively, an orthogonal I/O address space together with additional I/O instructions is needed. Finally, dedicated I/O ports like in conventional ASIC hardware blocks have to be used e.g. for handshake, start and stop signals. This interface mechanism can be supported by special instructions and/or by some kind of program flow or program interrupt controller.

4.3.9 Tightly-Coupled ASIP Accelerators

Accelerators are optimized dedicated hardware structures, which are typically able to perform a very limited set of computational tasks. Tightly-coupled ASIP accelerators can be viewed as elaborate functional units in the ASIP, which are integrated in the instruction set architecture. This integration is reflected by the interface and control mechanisms that are typically used. The interface between the ASIP core and the accelerator are either realized using specialized internal registers or simply the general purpose register file. On the other hand, the control of accelerators can be performed by using specialized ASIP instructions like in [15] and/or specialized registers similar to residual functional control like in [180]. The difference between a tightly coupled ASIP accelerator and an ordinary functional unit in the ASIP is the fact, that accelerators typically implement more complex sequences of operations. This requires a more complex internal structure often with an ASIP-independent controller.

Accelerators are typically used, if at least one of the following conditions is fulfilled:

- extremely high computational performance is required, which can not be satisfied by modification of ordinary functional units

- extremely high energy-efficiency is needed

In addition to the implementation of the already mentioned interface and control mechanisms of ASIP accelerators, the designer has the same degrees of freedom for their internal implementation than for dedicated hardware blocks: The most important decision is based on the trade-off between additional area consumption (which corresponds to the degree of parallelism in the accelerator implementation) and the additional computational performance. However, there are further important decisions that affect the overall implementation flexibility: Ideally, ASIP accelerators should only be used for tasks with a very low probability of late design changes. This strategy minimizes the risk of a chip re-design due to late design changes.

4.4 Critical Factors for Energy-Efficient ASIPs

The question arises, which of the design axes of Section 4.3 are most important in order to implement computationally optimized, energy-efficient ASIPs. The issue behind that question is, that there is no single ASIP application that represents all possible applications in the world[2]. This in turn makes it hard to deduce common properties and propose common guidelines. Each ASIP application has its own characteristics concerning typical operations, typical data transfer schemes, data rates and additional constraints. Obviously, the right question to ask is, how the designer can find the critical parameters and how to tune them in order to achieve a certain application-specific design goal. These questions related to the design flow will be answered in Chapter 5. Prior to tuning parameters it is important to understand the principal effects of important design decisions. This is the focus of the following subsections starting with the typically most important timing and performance constraints. Afterwards, the impact of ASIP modifications on energy and area consumption is discussed.

4.4.1 Timing and Computational Performance

Many high level ASIP design approaches like [17] or [70] use the abstraction of machine cycles as a metric to evaluate the result of a design modification. This approach does not consider the impact on low level timing (critical path T_{crit}) of synchronous logic using edge-triggered flip-flops [67]. The critical path T_{crit} of such a circuit determines the maximum operating clock frequency $f_{max} = 1/T_{crit}$.

The change of T_{crit} can be significant in some cases e.g. in [101], where the dramatic effect of small, incremental modifications of functional units, the control logic and the memory units on the maximum clock frequency is evaluated. An increase of up to 30% in the case of small changes in the ALU and an increase of up to 60% in the case of a modification in the branch unit emphasize the importance for early low level hardware estimations.

[2]This issue is in analogy to "the Ultimate Question of Life, the Universe and Everything. All we know about it is that the answer is 0b101010" [4].

Two design approaches are possible:

- the critical path T_{crit} is constrained by the ASIP system environment (typical low-end embedded application scenario)
- the ASIP is running (nearly) at the maximum speed f_{max} and the minimization of the total runtime $T_{run,min} = T_{crit}N_{cyc}$ of a task (which requires N_{cyc} processor cycles) is the optimization goal (high end application scenario)

In the first case, the critical path of the ASIP optimization is upper-bounded by the system clock frequency. During ASIP design, it has to be guaranteed that this constraint is not violated by any ASIP modification. This means that after each major or minor hardware modification, the hardware estimation design flow (refer to Section 5.3) has to be repeated in order to check this constraint. This methodology is in analogy to conventional HDL-based hardware design, where automatic synthesis has to be regularly performed after design modification to check low level constraints. In order to obtain a moderate design time, while exploring a sufficiently high number of different ASIP implementations, it is mandatory that this hardware design flow should be automated to a large extent.

In the second case, incremental ASIP changes are performed with the goal to minimize the total runtime $T_{run,min}$ of a task. This might be useful in the case of programmable accelerator chips (e. g. for high-end graphics applications like [58] or [106]), where high data throughput is a competitive advantage. In order to achieve this goal, the product $T_{crit}N_{cyc}$ has to be reduced. As previously mentioned, even small changes to the ASIP architecture that reduce N_{cyc} can lead to a significant increase in T_{crit}. To make matters worse, the reduction in N_{cyc} is typically strongly application-specific, thus, late design changes of the application might lead to suboptimal performance. An example for such a worst case is the scenario shown in [77], where the instruction set has been (over-) optimized for MD5 (*message digest*) encryption, which was actually harmful for a different algorithm (SHA - *Secure Hash Algorithm* [220]). Such a worst case represents the bound of flexibility and the risk of over-specialization of an ASIP implementation.

However, properly designed ASIPs typically take advantage of changes in the data path, without significantly affecting the critical path. This can be achieved by parallelization of computations using **parallel functional units** supported either by replicated decoders with an additional dispatcher like in Chapter 7.2 or by specialized instructions. The interconnection structure of a parallelized data path has to be designed carefully to avoid communication bottlenecks in large general purpose registers or large power and area consuming switch matrices like in [81]. On the other hand, approaches that emphasize the **chaining of operations** [262] risk to increase the critical path. If the increase of T_{crit} can be tolerated (e.g. because it does not violate the clock constraint of the system environment), it has to be (over-) compensated by a corresponding decrease of N_{cyc} in order to achieve a benefit for the total execution time. However, if the increase of T_{crit} can not be tolerated, **retiming** of logic can be performed and/or **additional pipeline registers/multi-cycle operations** can be introduced. Retiming (cf. Subsection 3.3.2), which can be manually or automatically performed, has the goal to balance the delays of combinational logic in different stages. Retiming is only possible, if the critical cyclic graph of logic contains at least two registers. Multi-cycle operations mixed with single cycle operations are obviously feasible, but they make the implementation less orthogonal and increase the verification effort. Introduction of additional pipeline stages also tends to increase the penalty for taken branch instructions and increases in turn N_{cyc} (refer to Section 7.1 for an example, where this has happened).

If the above-mentioned approaches to ASIP performance optimization fail to meet the constraints of an application, the implementation of a **tightly-coupled ASIP accelerator** is an option. This implementation, however, corresponds to a shift of the ASIP implementation towards more dedicated hardware, which has to be carefully considered in order to avoid an unnecessary decrease in the overall implementation flexibility. In many cases it is possible to use the accelerator for a limited subset of a runtime-critical task (e.g. a loop body like in [222]) which actually does not require a significant amount of flexibility.

If the application exposes a significant amount of data parallelism, it might be advantageous to implement parts of the ASIP as data parallel architecture. In this case, appropriately high **memory bandwidth** has to be provided in order to keep the functional units busy. One of the most important advantages of processing elements and memories on a single chip is the fact, that memory bandwidth is (theoretically) only bounded by the exploitable data parallelism of the application and the available silicon area (both for memories and functional units). For off-chip communication the chip pad limits are a major cost factor and obstacle to implement a high bandwidth interface. This fact naturally leads to a heterogeneous, non-hierarchical, but partially parallel memory architecture with small, fast scratch pad memories for intermediate values and larger (and possibly slower) main memories. The use of a memory hierarchy with level one, level two and main memory would also be an option to increase the bandwidth to main memory. In the case of large off-chip external main memory, this concept is needed in order to decrease the memory latency of each access. For typical cyclostationary DSP kernels with a limited amount of required data storage, however, the memory access schemes are regular and easily predictable, thus, the introduction of cache memories optimized for irregular (general purpose) access patterns is typically overhead. This is one of the main differences between ASIPs and general purpose digital signal processors like TI's TMS320C6x [246], which extensively uses such a cache hierarchy at the expense of a decreased energy-efficiency.

Finally, the **data I/O** for high performance ASIPs is a critical factor, because it can decrease the utilization of functional units, if the processor itself has to take care for it. One solution to this is a high speed DMA controller with exclusive access to parts of the ASIP memory. An even more elaborate scheme is the combination of conventional direct memory access controllers with a suitable double buffering scheme. Double buffering reserves two parts of the (shared) memory: one part is used by the DMA controller and the other part by the ASIP. After DMA and ASIP data processing has finished, a simple control logic exchanges the two parts of the memory in order to enable DMA access to the second part and ASIP access to the first part.

4.4.2 Energy Consumption

The effects of architectural changes on the energy consumption for a given computational task are more complicated than the above-mentioned effects on computational performance. This is mostly due to the statistical nature of power consumption, which is affected by data correlation of subsequent binary values on the nodes. This fact requires detailed tool-supported power analysis for the relevant operation scenarios.

The following discussion uses the abstraction of word-level hardware operators, which is the natural level of abstraction for HDL-based hardware design. In analogy to [54] where the term *intrinsic computational efficiency of silicon* has been introduced the following terms are defined for simplification purposes:

- In case of a full match between application and architecture, each hardware operator (like e.g. an adder or multiplier) is contributing a useful calculation to the overall computational task. For a large set of stimuli, the average energy consumption of this ideal architecture can be calculated, which shall be called **Intrinsic Computational Energy** E_i. In the case of an adder this energy is a function of the technology, the operating conditions (like supply voltage and temperature) and of the adder implementation (energy evaluations of different adder implementations can be found in e.g. [190] or [196]).
- The difference between the overall energy E_{tot} of a synchronous, instruction set oriented processor (including all the memories that are needed to process the task under consideration) and the intrinsic computational energy E_i is called **Overhead Energy** E_{ovhd} is needed for control logic (including the program memory), data memories, data transfers between processor units (routing energy), additional spurious transitions (other than those that are already included in the intrinsic energy of the operators), and the energy consumed in the clock tree and in the registers.

The intrinsic energy E_i is the lower bound in energy consumption that is needed by an ideal (dedicated) hardware data path without any additional overhead energy due to glitches or clock networks. Even optimized real hardware needs either energy for a clock network and registers or - in the case of a pure combinational network - it needs additional energy for unavoidable spurious transitions (glitches) due to the signal timing slack of intermediate results.

The percentage of overhead energy to overall energy can be significant: [223] reports an overhead energy of at least 64% and [231] and [249] estimate overhead energies of at least 79% and 70% respectively. In [124] a range between at least 61% for embedded processors and at least 72% for high end processors is reported. The fact that the overhead energy for a processor is of this order, agrees with the results of the case study in Section 7.1 of this thesis.

ASIP optimization in order to lower the energy consumption has to decrease the overhead energy. One solution to achieve this, is to decrease the runtime of the given task by application-specific data path optimizations and/or by an optimized software implementation. Figure 4.6 illustrates this effect, which relies on the assumption that the overhead power is (nearly) unaffected by the optimization. Obviously, not all of the architectural changes that have been described in the previous subsection are able to meet this assumption:

- **Parallel functional units** or **ASIP accelerators** that represent a close match to the application's control data flow graph are a typical example for efficient low-power data path optimization. The principle of this technique is to increase the rate of operations without increasing the rate of instructions. This implicitly requires dedicated application-specific instructions to support the increased parallelism in the data path. This optimization leads to architectures that are beyond the typical SIMD class of processors, because the data path is not restricted to perform the same computations on a set of data. A single highly optimized ASIP instruction can rather trigger a number of arbitrary data processing operations.

- If **chaining of operations** is possible without violating the time constraints and without introducing additional registers, this modification also tends to increase the energy-efficiency. It also typically requires adding one or more instruction to the ASIP instruction set, which leaves the overhead energy nearly unchanged (at least in a simple single issue processor). Unfortunately, if new interconnections between e.g. the general purpose register and the chained operators have to be introduced, the size of the interconnection networks increases, which in turn increases the overall data routing energy for any data transfer on the modified intercon-

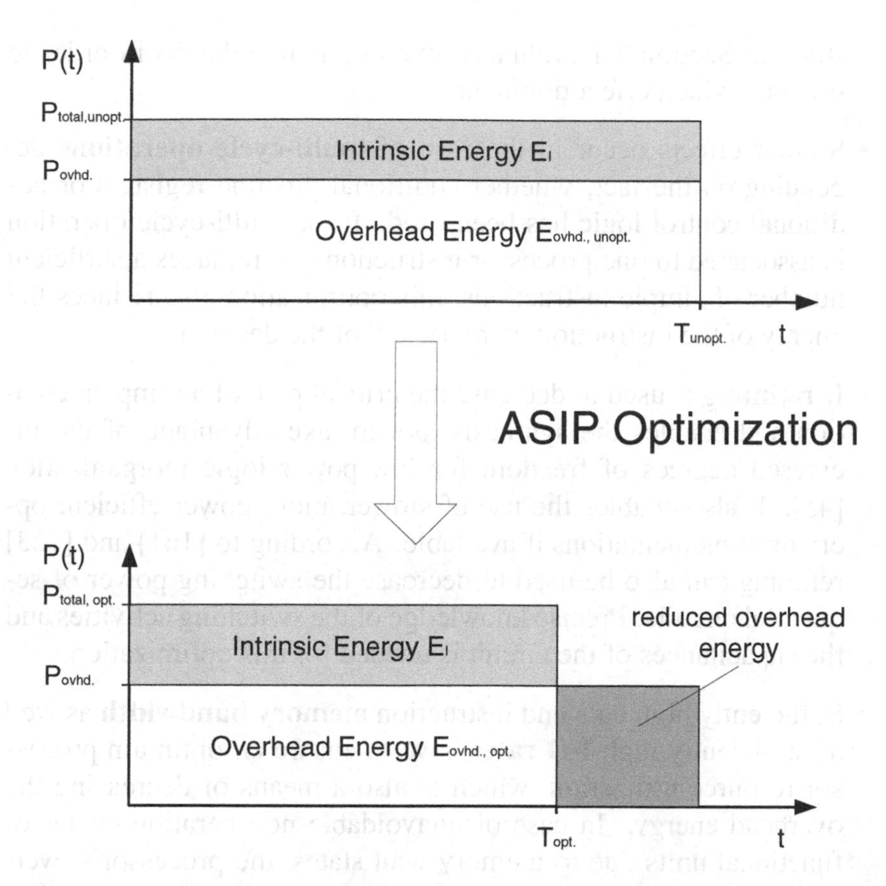

Figure 4.6: Principle of Energy Reduction with Optimized ASIP Architecture

nections. The overall effect of this modification has to be thoroughly evaluated in each case in order to find out, if the modification was successful.

- **Additional pipeline registers** to increase the pipeline depth have several effects: Firstly, pipeline register reduce the spurious activity by resetting the signal slack (the difference between the earliest and the latest signal arrival event) to nearly zero. Secondly, additional pipeline registers need a larger clock tree and the registers themselves require additional energy due to clock activity. Finally, additional pipelines possibly result in larger branch penalties and more complicated logic to detect and resolve hazards. The case study in Section 7.1 evaluates several pipeline depths in order to find out, which effect dominates.

- Similar effects occur in the case of **multi-cycle operations** depending on the fact, whether additional pipeline registers or additional control logic has been used. If the multi-cycle operation is associated to one processor instruction that replaces a sufficient number of simple instructions, this optimization also reduces the energy of the instruction memory and of the decoder.
- If **retiming** is used to decrease the critical path of an implementation and enables the synthesis tool to take advantage of the increased degrees of freedom for low-power logic reorganization [45]. It also enables the use of slower, more power efficient operator implementations if available. According to [161] and [183] retiming can also be used to decrease the switching power of sequential circuits. Precise knowledge of the switching activities and the capacitances of the circuit is needed for this optimization.
- Sufficiently high data and instruction **memory bandwidth** as well as sufficiently high **I/O rates** have to ensure an optimum processor resource utilization, which is also a means of decreasing the overhead energy. In case of unavoidable no-operation cycles of functional units due to memory wait states, the processor's overhead energy should be reduced as far as possible (e.g. by clock gating and sleep modes).

Apart from reducing the overhead energy by reducing the runtime of a task, the overhead power P_{ovhd} can be reduced directly e.g. by optimizing one of the following:

- high coding density reducing the size and the energy consumption of the instruction memory (this corresponds to a high average ratio of executed operations per instruction bit – this metric is also implicitly optimized by SIMD extensions, operator chaining and ASIP accelerators)
- optimized instruction encoding (refer to Subsection 7.1.3) reducing the energy consumption within the instruction memory or alternatively, the switching energy of an external instruction bus

- reduced data memory accesses by software optimization together with a sufficiently high number of local registers
- optimized data encoding for high capacitive nodes
- application-specific (limited) interconnections between functional units or within an ASIP accelerator to decrease the data routing energy by decreasing the effective switched capacitances
- guard logic to reduce/avoid spurious transitions in combinational logic (refer to Subsection 3.3.2 for further explanations)
- clock gating in order to shut down idle parts of the clock tree and to reduce the energy needed in the connected flip-flops

All of these direct optimizations lead to a more energy-efficient implementation typically without a negative impact on computational performance or silicon area.

If the ASIC technology vendors were going to support characterized cell libraries for a larger voltage range than today, this would also enable the designer to reduce power by taking advantage of a reduction in clock frequency for less computationally demanding tasks together with aggressive voltage scaling.

4.4.3 Implementation Area

Due to the fact that technology scaling continues to follow Moore's law [184] so far without a perceptible deceleration, area consumption is gradually getting less important. Nevertheless, area is currently still a costly resource, which affects the unit price of an ASIC and has to be minimized in order to increase the profit margin for high volumes.

Area consumption can also be an issue due to the following reasoning: an increase in area increases the average length of interconnections (which can be estimated from wire load models [93] [256] used for synthesis), which in turn decreases the maximum clock frequency. Thus, excessive area increase has to be avoided in order to avoid performance degradation. To make matters worse, an increase of the average interconnection length also increases the interconnection capacitance resulting in a higher switching power. Long interconnection delays can be avoided by using local communication rather than global communication. This means, that the interconnections between hardware blocks should be reduced to a minimum in order to save area and energy and to enable high computational performance. Properly designed ASIPs follow that guideline and use local communication between adjacent functional units.

It should be emphasized that ASIPs are inherently saving area, because they represent resource shared architectures[3] compared to typical dedicated hardware, which often exposes more parallel processing structures. From a hardware perspective, ASIP design can be viewed as a design methodology in order to implement resource sharing by extracting and implementing common operation patterns from a given control data flow graph representation of an application.

4.5 Concluding Remarks

This chapter has defined important terms for ASIP design as well as typical ASIP applications. Furthermore, the design space for ASIPs has been characterized with the goal to provide well-defined degrees of freedom for the ASIP designer. This characterization enables both the ASIP designers together with the compiler designers to decide, whether specific architectural features of an ASIP are needed and if compiler support for these features is possible. This design process is an important aspect for the design efficiency of ASIPs and is commonly referred to as compiler/processor co-design [284]. Finally, this chapter has qualitatively discussed the effects of high level ASIP design decisions on the computational performance, the energy consumption and the implementation area. This knowledge is needed in order to successfully apply the design methodology described in the following chapter to real world applications.

[3] Area can however be an issue for massively data or instruction parallel architectures or for ASIPs with parallelized accelerator extensions.

Chapter 5

The ASIP Design Flow

The design of ASIPs represents a hardware/software codesign task, because hardware and software related expertise is needed in order to get optimum results. This codesign problem can be considered as a complex optimization problem in the multi-dimensional ASIP design space, which has been defined in Chapter 4 and, additionally, also as a software optimization problem. It is the primary goal of most ASIPs design to find a programmable architecture that meets the performance constraints of an application and that consumes a minimum in area and energy consumption. Moreover, this architecture should be sufficiently flexible to cope with late design changes due to evolving standards or incorrect specifications.

This chapter provides a complete description of the proposed ASIP design flow[1], which starts with a behavioral description of the algorithm and ends with the optimized ASIP hard- and software implementation including design tools and documentation. One important feature of the proposed ASIP design flow is the fact, that the ASIP hard- and software is in the iteration loop. This enables the designer to jointly optimize performance, silicon area and energy in order to get a feasible implementation in a short amount of design time.

An overview of the entire ASIP design flow is depicted in Figure 5.1 starting with the behavioral high level language (HLL) specification of the application. The ASIP design tasks in Figure 5.1 represent a top-down design approach, which enables the designer to cope with complexity by using several abstractions. During the initial application profiling, the abstraction of high level language statements and operators is used in order to collect execution statistics of the application. For the initial instruction set architecture definition, the designer can use a cycle-true description of the instruction behavior, which neglects details

[1] The examples in this chapter focus on ASIP performance enhancements rather than energy optimizations, which simplifies the discussion. Chapter 7 extends this focus by covering the energy consumption and the design time of ASIPs using two real-world examples.

of the low-level hardware implementation in the first place. Afterwards, the implementation of the hardware RTL description uses the same high level operators, neglecting the specific logic implementation, which is added later on either by explicit definition or by logic synthesis[2]. These abstractions imply, that precise low level estimations can lead to iterations in the design flow resulting in changes of higher level design decisions.

The design flow in Figure 5.1 starts with an arbitrary application, which might contain parts suitable for an ASIP software implementation and other parts suitable for a more dedicated hardware implementation. In order to identify this partitioning and perform the mapping to hard- and software, several design tasks are needed that are beyond the definition of an instruction set architecture. Nevertheless, these design tasks are important in order to realize an optimum implementation in any case.

The described ASIP design flow is adaptable to the needs of typical applications, which is demonstrated by examples. The examples in this chapter should not be regarded as complete case studies, but rather serve as vehicles in order to illustrate important design decisions. The selected example applications have been chosen as a representative subset of possible ASIP applications in order to cover relevant DSP applications.

This chapter is organized as follows: In Section 5.1 the example applications and example kernels are briefly introduced. Section 5.2 depicts the application profiling and partitioning tasks, which are needed prior to the actual ASIP design tasks described in Section 5.3. The examples in Section 5.2 and Section 5.3 are made standing out using serif font.

5.1 Example Applications

In this section the ASIP applications are presented that will be used as examples in the next section. These examples have been selected in order to cover a reasonably large part of the ASIP design space concerning

[2] Alternatively, for high speed arithmetic either custom designs or tools like Synopsys' Module Compiler [236] can be used.

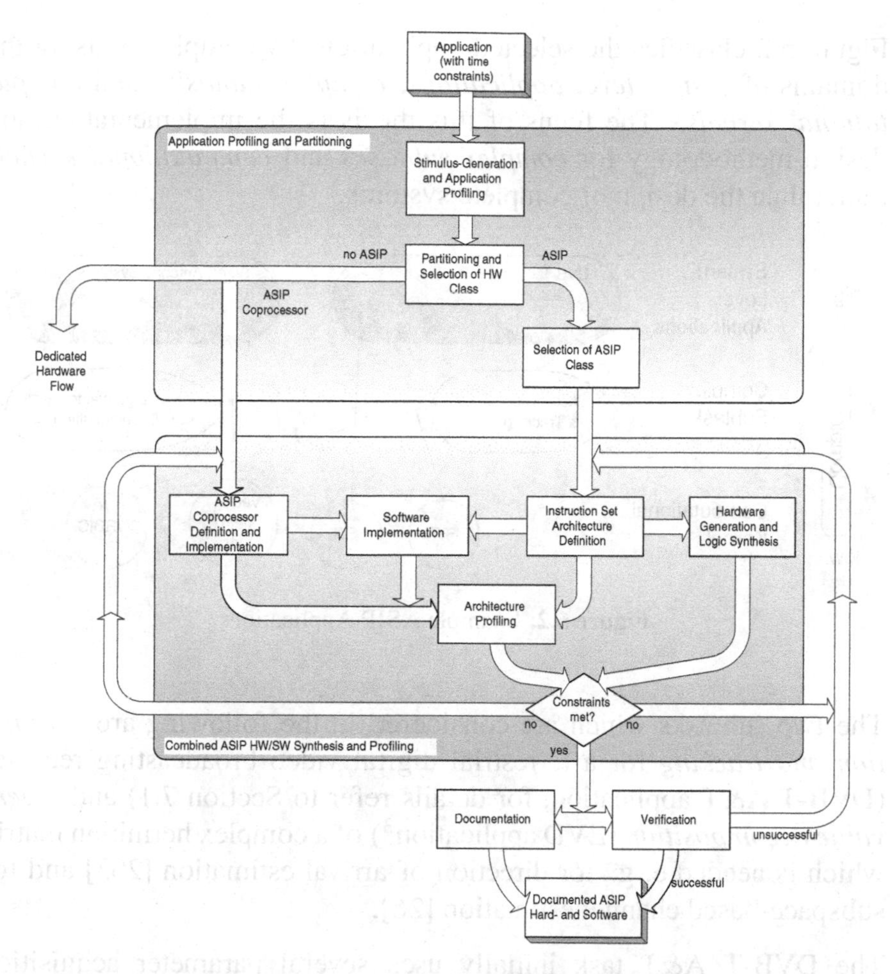

Figure 5.1: Overview of ASIP Design Tasks

- complexity of the task (in code lines of a high level language (HLL) description)
- control flow vs. data flow orientation
- cyclostationary vs. non-cyclostationary data processing
- high vs. low data locality
- high vs. low data rate and computational requirements
- different operator granularity

Figure 5.2 classifies the selected applications by complexity using the domains of *system level applications*, *complex subtasks*, and *computational kernels*. The focus of this thesis is the implementation and design methodology for *complex subtasks* and *computational kernels* rather than the design of complete systems.

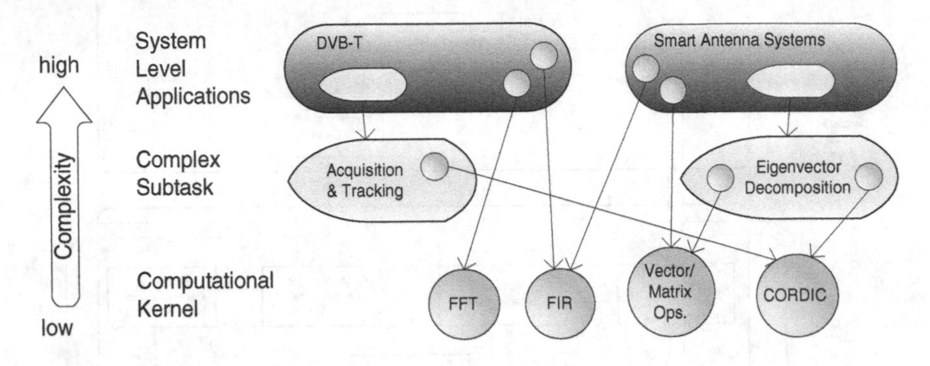

Figure 5.2: Example ASIP Applications

The two subtasks which are considered in the following are *acquisition and tracking* for a terrestrial digital video broadcasting receiver (DVB-T A&T application, for details refer to Section 7.1) and *eigenvalue decomposition* (EVD application[3]) of a complex hermitian matrix which is needed e. g. for direction of arrival estimation [203] and for subspace-based channel estimation [26].

The DVB-T A&T task initially uses several parameter acquisition phases that expose non-cyclostationary processing and finally enters a theoretically continuous parameter tracking which is largely cyclostationary. Due to the huge number of control parameters of the DVB-T A&T tasks, this application represents a mixed control/data flow oriented application. On the other hand, the EVD uses cyclostationary processing, because it represents a data flow oriented architecture that is based on the granularity of matrix block processing. Parts of the DVB-T A&T task require very high computational performance, which are directly determined by the DVB-T transmission time frames, whereas other parts only affect the acquisition time of the system. On the other hand, the EVD for a direction-of-arrival (DOA) estimation requires high

[3]The EVD for the example applications use a matrix size of 10x10.

computational performance, because this impacts the number of supported mobile users for a mobile base station.

The different computational kernels that have been selected for illustration purposes are *finite impulse response filters (FIR)*, *fast fourier transformation (FFT)*, and *coordinate rotation digital computer computations (CORDIC)*. Descriptions and listings of the high level language implementations[4] which have been used as behavioral descriptions for these kernels can be found in Appendix B.

The properties of the selected applications are described in Table 5.1. Examples for constraints concerning the data rate are given later on in this chapter. It has to be mentioned that several properties in Table 5.1 depend on the software implementation. For instance, the data locality of the FIR depends on the implementation of the delay line for the input samples: If this delay line is realized with explicit memory move operations, the data locality is medium, whereas if a circular buffer is used, the data locality is high. Another example is the SW implementation dependent granularity of operators, which can be refined for any application to the granularity of standard word-level or bit-level operators.

	Data Locality	Granularity of Operators	Complexity (code lines, states)	Control/ Data Flow Domination	Cyclo-stationary Processing
FFT	low	complex scalar	medium	data	yes
FIR	medium to high	real scalar	medium	data	yes
CORDIC	high	real scalar	medium	mixed	yes
EVD	medium	complex scal./vect.	medium/ high	mixed contr./data	largely
DVB-A&T	high	real scalar	high	mixed contr./data	during tracking

Table 5.1: Properties of the Different Selected Example Applications

[4]The key parameters for these kernels are: 64 taps for the FIR, 8192 point for FFT, and 24 iterations for the CORDIC task.

5.2 Application Profiling and Partitioning

The proposed design flow of Figure 5.1 starts with a HLL description of the algorithm, which has to be profiled in order to identify critical parts. Criticality on this level of abstraction refers to parts of the algorithm that require high computational performance and/or high memory bandwidth. With these results, the partitioning into parts of the application that will later on be mapped to ASIPs, ASIP coprocessors or dedicated hardware can be performed.

5.2.1 Stimulus Generation for Application Profiling

This subsection describes the requirements and issues of stimulus generation for application profiling. Typically, either **stimulus generation** from scratch or **stimulus reuse and extraction** of already available stimuli by the system simulations is required to expose the performance critical parts of an application. This profiling stimulus generation design task does not have to provide full simulation coverage of all arithmetic and logical functionalities of the reference and of all internal states (like the stimulus generation for verification); it rather has to produce the worst case runtime[5] of the application to obtain realistic profiling results for real-time applications[6]. A comparable stimulus generation task has to be performed after an initial ASIP implementation is available in order to generate worst case stimuli for architecture profiling.

Example: Due to the complexity and configurability of the DVB-T A&T application, profiling stimuli could not be generated by the system simulation. Table 5.2 depicts the complexity of all the different DVB-T A&T tasks in number of C code lines and number of I/O parameters as well as the complexity of the SW testbenches. Each of the tasks Pre-FFT-, Post-FFT-Acquisition, and Post-FFT-Tracking has about 100 internal state variables: The control flow in the application is a function of a smaller subset of these state variables. This subset of state variables has been identified and suitable testbenches have been manually written in order to execute the critical path in the software for worst case runtime. For this complex application with many boundary cases[7],

[5]Worst case runtime can be excited by requiring the maximum number of operations/data transfers per time unit of a latter implementation.

[6]For non-real-time applications, the typical runtime can also be used as a metric for this design task.

[7]These application typically use deeply nested HLL control statements like e.g. *if or switch.*

a divide and conquer approach has been used, which partitions the application and the testbench into smaller pieces, that can be independently profiled[8].

The initial effort to create these software testbenches adds up to more than two man weeks for the DVB-T A&T application, which corresponds to more than 5% of the total design time. These testbenches in modified form have been reused later on in the design flow for architecture profiling and for the verification between the software running on the instruction set simulator and the reference software.

Task	Behavioral Description (# of C Code Lines)	# of I/O Parameters	Testbench (# of C Code Lines)	SW Testbench Design Time (# of Man Days)
Reset	44	-	5	< 0.1
Pre-FFT-Acqu.	359	10/3	275	5
Post-FFT-Acqu.	330	10/2	283	4
Post-FFT-Tracking	397	4/3	278	7
PEQ Estimation	60	3/1	(incl. in Tracking)	-

Table 5.2: Design Effort for SW Testbenches (DVB-T A&T)

5.2.2 Application Profiling

The purpose of application profiling is to locate the performance critical parts within an application using profiling stimuli for worst case scenarios. Application profiling ideally should be target-architecture-independent: Profiling at the abstraction level of operators and memory accesses with a certain user-defined data granularity should be preferred over measuring the instruction count of a real implementation. Obviously, for HLL-code-based applications, it makes sense to use typical HLL code operators for profilingand measure the memory accesses to primitive data structures (like e.g. one integer element in an integer array).

[8]For this purpose, the application has been temporarily modified to enhance the controllability.

In order to perform this profiling task (or an approximation of this task), two basic approaches are possible: operator-based profiling or HLL line-based profiling. Table 5.3 shows a comparison of these approaches.

Approach	HLL Operator-Based	HLL Line-Based
Tool/Methodology	instrumented HLL code [56]	e. g. gcov [229]
Design Effort	high	low
Execution Speed	lower	high
Precision of Results	high	low

Table 5.3: Comparison of Application Profiling Approaches

The line-based approach can be readily performed with a coverage tool like e.g. gcov [229] with virtually no additional design effort. Actually, a coverage tool gcov is supposed to evaluate line coverage information, but the output of this tool can obviously also be used for profiling purposes. This approach enables the designer to obtain an overview of critical loops in the application within a very short amount of time. Unfortunately, the accuracy of the results is poor, because the number of high level language operators and memory accesses on each HLL line varies.

On the other hand, true operator-based profiling using an instrumented HLL code requires to add additional profiling statements in the code, which leads to a significant increase in design time. More advanced techniques for this design task use overloaded operator instances, which update the profiling information as a side effect. For this thesis, however, a different application profiling approach is proposed. This proposed profiling methodology uses the concept of a so-called *profiling processor* as a virtual target architecture together with an optimizing HLL compiler. The HLL application is mapped to this profiling processor and the instruction/operator count as well as the memory accesses are evaluated by an instruction set simulator. It may be argued that this concept violates the target architecture independence of the application profiling task, because a real instruction set architecture is used as a target for profiling. This objection is partially true, but the advantages of the proposed methodology outweigh the demerits:

- the profiling processor can be taken out of a processor template library (PTL[9]) (together with the HLL compiler and the simulation tools), which enables a *complete automation* of the application profiling task
- additional information (apart from operator count and memory accesses) can be obtained by this approach, like data locality, branch frequency etc.
- the results are accurate, in the sense, that the measured instruction count on this profiling processor could be implemented with a real instance of this processor
- exactly the same methodology can be used later on for architecture profiling by taking the processor template (out of the PTL) for the processor class (cf. Subsection 5.2.4) that optimally matches the application and by using this processor template as a starting point for design space exploration

The most simple profiling processor implements a so-called *basic instruction set* with two operand instructions (register/register or immediate/register), which has been described in [91] and which is reprinted in Table 5.4. Two separate flat address spaces for I/O and data memory can be accessed by register indirect addressing with displacement.

This instruction set is only a subset of typical HLL operators, because operators like division and modulo operations are usually very expensive in terms of either silicon area, energy or latency and, consequently, have been omitted in this instruction set. Obviously, these or other operators can be added, if an application extensively needs them. Furthermore, explicit I/O instructions have been provided in order to profile the input/output behavior of the algorithm[10].

This profiling processor uses a one stage pipeline organization, which avoids forwarding paths and corresponds to an instruction true abstraction from a real pipelined implementation. The architecture does not support instruction level parallelism, but rather provides a scalable general purpose register file optionally with a cache memory hierarchy in

[9]Implementations of the software tools and the hardware for different processor classes (processor classes can include the examples of Subsection 5.2.4 or the classes defined in Section 4.3) can also be stored in this PTL to provide a starting point for the design space exploration and implementation.

[10]These I/O instructions can be supported by compiler-known-functions.

Type of Instruction	Instruction Mnemonic	Description
Load/Store	RDIO	read I/O data
	WRIO	write I/O data
	RDM	read data memory
	WRM	write data memory
arithmetic	ABS	absolute value
	ADD/ADDI	addition
	MOV/MOVI	data move
	MULU/MULS	signed/unsigned multiplication
	SHL/SHLI	arith./logic. shift left
	SRA/SRAI	arith. shift right
	SUB/SUBI	subtraction
logic	AND/ANDI	bitwise AND
	OR/ORI	bitwise OR
	SRL/SRLI	logic. shift right
	XOR/XORI	bitwise XOR
control	CMP/CMPI	compare/set status
	BRA	uncond. branch
	BSR	branch to subroutine
	BEQ/BNE	branch if equal/not equal
	BLT/BLE	branch if less than/less or equal
	BGT/BGE	branch if greater than/ greater or equal
	END	exception/transition to idle mode
	RTS	return from subroutine

Table 5.4: Basic Instruction Set for Profiling

order to measure the data locality of data flow intensive algorithms. The HLL compiler for this profiling processor has been implemented with the COSY Compiler Development System [3].

Example: Figure 5.3 shows the result of HLL line-based profiling for the DVB-T A&T application, where loop kernels with a significant number of iterations can be clearly identified. This profiling run uses realistic stimuli, which reflect the different states of the processor, namely a relative short period for the acquisition of different parameters and afterwards the (theoretically continuous) tracking operation. The large range in execution frequency is the reason that a logarithmic scale on the vertical axis in Figure 5.3 has been chosen.

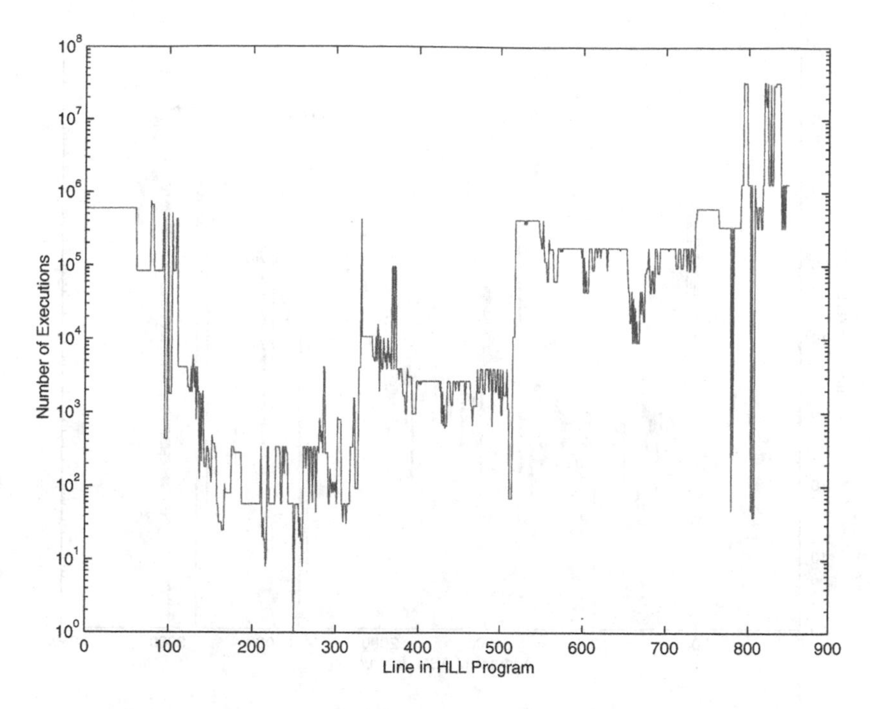

Figure 5.3: HLL Line Coverage Profile (DVB-T A&T example)

The above described profiling methodology using the profiling processor has also been applied to the DVB-T A&T application. Figure 5.4 shows the profiling results for the assembler implementation with the same stimuli that have been used for Figure 5.3. The similarity of the two visualized profiling graphs is obvious, which is due to a close match between the compiler generated assembly implementation and the reference HLL implementation. However, the scale of the horizontal axis clearly shows, that several assembly instructions per line of HLL code have been executed. Provided that it is possible to implement a profiling processor that achieves one instruction execution per cycle for a given clock period constraint, the vertical axis in Figure 5.4 corresponds to clock cycles in the system. In this case, the (non-logarithmic) area under the graph for a certain address range in the program memory is proportional to the runtime which is spend in this part of the program.

In order to assess the performance criticality of the application, quantitative data are needed, namely the ratio between worst case runtime of the profiling implementation and the maximum runtime constraint of the application. Obviously, this ratio should be smaller than 1.0 in order to obtain a feasible implementation. Table 5.5 depicts the performance evaluation results for the example applications under the assumption that the profiling processor has

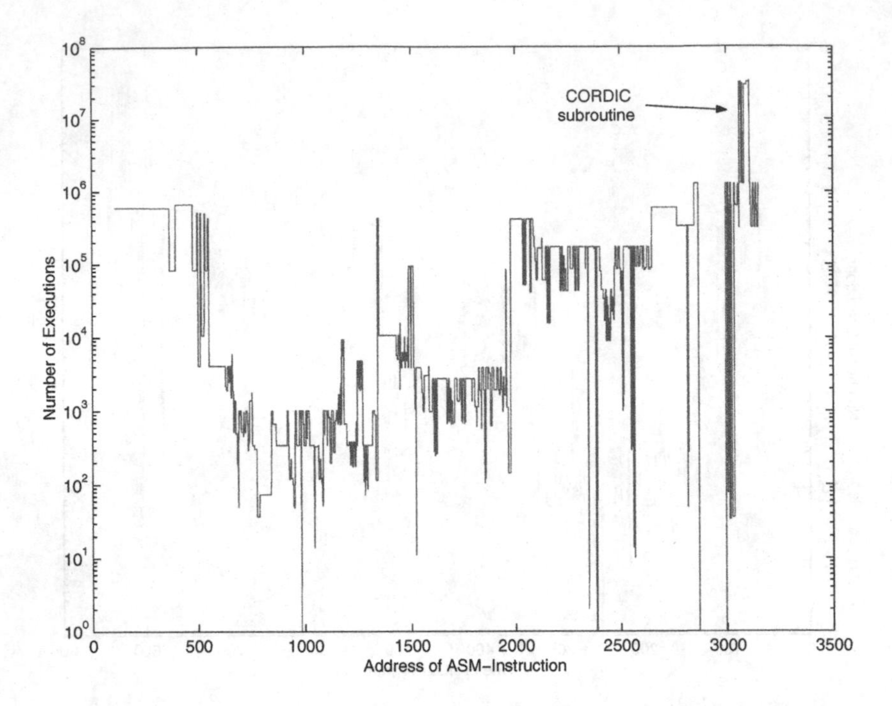

Figure 5.4: Assembler Line Coverage Profile (DVB-T A&T profiling impl.)

32 general purpose registers and is running at the system clock frequency of the DVB-T system. The cycle constraints of all the computational tasks are violated in Table 5.5, which motivates optimizations of the processor architecture implementation.

Application	Worst Case Cycle Count N_{wc} (on Profiling Processor)	Cycle Count Constraint N_{max}	Ratio $\frac{N_{wc}}{N_{max}}$
Pre-FFT-Acquisition	7077	4096	1.73
Post-FFT-Acquisition	5661	4096	1.38
Post-FFT-Tracking	6208	1024	6.06
PEQ Phase Estimation	1353	192	7.05

Table 5.5: Cycle Count of Profiling Implementation vs. Max. Cycle Constraints

For the sake of conciseness we limit the following discussion to the most critical DVB-T A&T tasks. After further examination of the Post-FFT-Tracking and PEQ Phase Estimation task, it was found, that these tasks make intensive use of a common CORDIC subroutine, which has been marked in Figure 5.4. Figure 5.5 illustrates the significant runtime, which is consumed in the CORDIC subtask with respect to the total runtime in each case. Consequently, optimization of the CORDIC subtask (possibly among other tasks) is needed in order to meet the cycle constraints for these critical computational tasks.

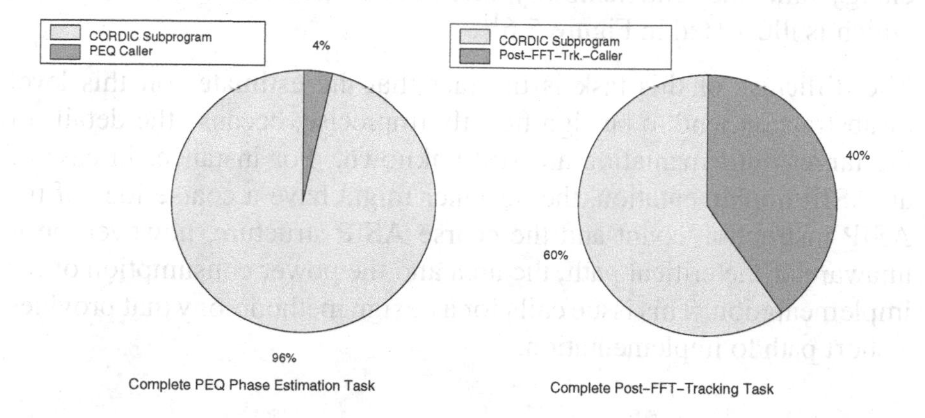

Figure 5.5: Percentage of Runtime used for the CORDIC Subtask

5.2.3 HW/SW Partitioning

HW/SW partitioning is a prerequisite to the actual ASIP design flow, which has to make sure that an instruction set oriented ASIP is a reasonable implementation for a given application or whether parts of the application are better mapped to coprocessors or dedicated hardware blocks. This decision is important for the energy-efficiency of the system, because the flexibility of ASIPs implies a higher energy consumption as demonstrated in Section 7.1. The logical consequence of this fact is that the high flexibility of ASIPs should only be used for tasks that require and take advantage of it. On the other hand, tasks with high computational requirements that do not need much flexibility should rather be mapped to a dedicated structure in order to take advantage of the higher energy-efficiency.

Partitioning of the application into blocks of a reasonable size is followed by the **selection of the hardware class** which maps each of these blocks to either a software implementation (e.g. ASIP, microcontroller, general purpose processor), to a tightly coupled ASIP coprocessor or to dedicated hardware. This selection task is primarily controlled by the estimated performance of the target hardware. For many tasks with sufficiently low time constraints, however, both hardware and software implementations are possible. In such a case, additional parameters like energy-efficiency and flexibility have to be considered for this selection, which is illustrated in Figure 5.6[11].

The difficulty of this task is the fact that the estimates on this level of abstraction tend to be significantly imprecise, because the details of the target implementation are still unknown. For instance, in case of an ASIP implementation, the designer might have a coarse idea of the ASIP instruction count and the coarse ASIP structure, however, he is unaware of the critical path, the area and the power consumption of the implementation. This issue calls for a design methodology that provides a short path to implementation.

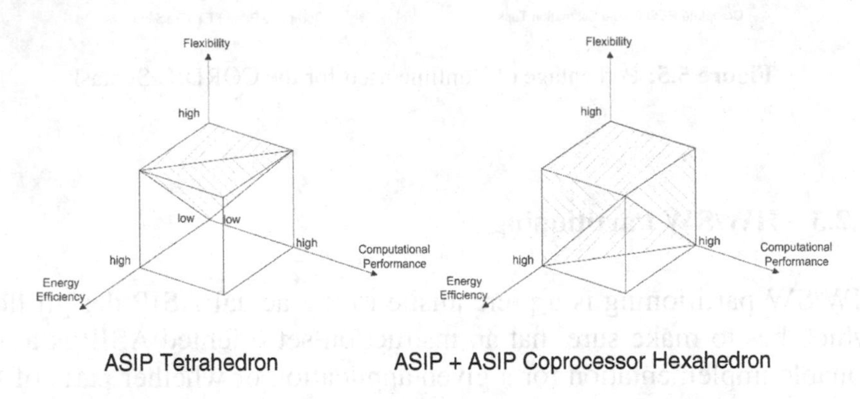

Figure 5.6: Design Space for ASIPs and ASIP Coprocessors

Example: For the above-mentioned CORDIC subroutine of the DVB-T A&T application, the selection of a hardware or a software implementation is not straightforward. On the one hand, the CORDIC requires high computational performance, which could be efficiently mapped to an energy-efficient ASIP coprocessor [86]. On the other hand, this coprocessor needs additional hardware resources, like shifters, adders and memories if implemented as a separate entity. Most of these resources are needed in the ASIP anyway and can therefore be resource shared. Furthermore, in the case of late design changes after silicon fabrication, the dedicated coprocessor needs a full redesign of the chip, whereas a software implementation typically needs only a redesign of the

[11] The remaining tetrahedron of the cube which is omitted in Figure 5.6 is the design space of dedicated hardware

program[12]. For the sake of increased flexibility, the CORDIC task in the DVB-T A&T application has been mapped to an optimized software implementation. Finally, in case of the EVD which requires higher computational performance than the DVB-T A&T application the CORDIC has been mapped to an ASIP accelerator.

5.2.4 ASIP Class Selection

The design space for ASIPs has already been described in Chapter 4. HLL support for complex applications is typically indispensable to avoid error-prone and tedious assembler coding. Provided that a HLL compiler or a compiler design environment (e.g. Cosy [3], Chess [156] or RECORD [165]) is required and available during the ASIP design, the compiler-supported subset of ASIP classes within this large design space has to be identified as a starting point for the following selection. Alternatively, in the absence of a compiler, the ASIP class has to be selected in order to facilitate manual assembler programming. Nevertheless, this is only possible for less complex applications in order to avoid tedious and error prone programming design tasks.

The task of finding the processor instruction set architecture (ISA) class that represents the best match to the application is of paramount importance for the design efficiency. This selection also affects the verification effort of the final hard- and software, which represents a major part of the overall design time according to Appendix F.

In the following discussion, several examples serve as illustration for the selection of important parameters for a suitable processor class. This discussion is not yet a complete commitment to a specific ISA. It rather determines a good starting point for the following ASIP optimization tasks.

Non-Parallel vs. Instruction/Data Level Parallel ASIP: A certain task can be implemented using a scalar (single instruction issue) ASIP, provided that the cycle count of the software implementation for the profiling processor in Subsection 5.2.2 is smaller

[12] In case of the DVB-T A&T application, this redesign of the program requires a redesign of the instruction ROM masks for the chip fabrication at reduced costs compared to a full chip redesign. For other applications, which use on-chip RAMs as instruction memories, a redesign of the ASIP software does not affect the chip fabrication costs.

than the cycle constraint with a certain cycle safety margin that accounts for the overhead of a real processor implementation. If this condition is not fulfilled, there are two options: either the ASIP has to be implemented as a classical data or instruction level parallel architecture e.g. using a SIMD or a VLIW implementation. Alternatively, the designer has to optimize the scalar ASIP instruction set by implementing additional optimized instructions that are able to perform several frequently needed operations in parallel during one clock cycle, which results in a cycle count reduction. This kind of optimization is a typical ASIP optimization, which also enhances the energy-efficiency of the implementation (refer to Section 7.1 for details).

Example: The choice between VLIW ASIP or optimized non-parallel ASIP significantly affects the design time and the area/energy-efficiency of the implementation. For the DVB-T A&T application, a scalar, single issue processor implementation has been chosen, in order to reduce the design time and to obtain the best possible energy- and area-efficiency. This choice requires optimization of the implementation using specialized instructions according to the results of application profiling. For the EVD an architecture that represents a mixture between a pure scalar and a SIMD architecture has been selected in order to speed up the processing of vector and matrix operations.

Organization/Access of Storage Elements: The memory and ASIP-internal register structure has to be organized in order to speed up the common and critical tasks of the application. This requires small and fast scratch-pad memories together with reasonably sized internal registers and register files. For applications with a high ratio $R_{l/s}$ of load/store-operations $N_{l/s}$ to the total number of executed operations N_{tot} the use of a pure load/store architecture with a central register file is disadvantageous. In such a case a memory-memory architecture or a heterogeneous architecture (which uses memory-memory instructions together with load/store instructions) is better in order to reduce the number of explicit load/store instructions, which need energy in the fetch/decode stage and result in a large footprint in the instruction memory. Furthermore, the significant energy overhead of writing and reading the general purpose register can be avoided.

Example: Table 5.6 shows the frequency of load/store instructions of the profiling implementation for the considered example applications. The

DVB-T A&T application can be readily implemented using a reasonably small, flat data memory in combination with a load/store architecture due to the small ratio $R_{l/s}$. However, the FFT has a large ratio $R_{l/s}$, which suggests a memory-memory architecture or alternatively, a heterogeneous architecture with optimized memory-memory instructions like in [138]. A more detailed application analysis for the EVD reveals, that it is indeed possible to take advantage of a large general purpose register file together with additional address registers in order to exploit data locality, which also suggest a load/store architecture. For the current FIR implementation, a memory-memory architecture is one option. However, a small change in the FIR high level description[13] using a circular buffer for the delay line can avoid about 50% of the memory accesses, which makes a load/store architecture also a feasible alternative.

Application	Percentage[14] of Load/Store Operations ($R_{l/s} \cdot 100\%$)
DVB-T A&T	3.5%
CORDIC	4.8%
EVD	17.4%
FIR	28.7%
FFT	36.8%

Table 5.6: Percentage of Load/Store Instructions (Profiling Implementation)

Pipeline: Depending on the clock period constraint of the system, the depth of the pipeline has to be chosen in order to meet this constraint. For a high operating frequency a long pipeline is necessary. On the other hand, a longer pipeline tends to expose a larger branch penalty for taken branches[15]. This is especially disadvantageous for tasks with a large ratio R_{branch} of taken branch operations N_{branch} to the total number of executed operations N_{tot}.

Example: In Table 5.7 the percentage of taken branch operations during program execution of the example applications is depicted. For the CORDIC, the FFT and the EVD this percentage is negligible, because a more thorough investigation of these applications reveals, that these branches are mostly used to implement loops which can be replaced by zero-overhead loop instructions. On the other hand, for the DVB-T

[13] The current description has been taken from the DSPstone kernel collection [285]

[14] For this analysis a general purpose register file with 32 general purpose registers has been used.

[15] This penalty model assumes a predict-untaken branch execution scheme [107] without executing instructions in the branch delay slots.

A&T application the significant number of taken branches reflects the property of a typical mixed control/data flow application. This high percentage motivates the use of a short pipeline in order to mitigate the overall branch penalty.

Granularity of Functional Units: For a good match between architecture and application, the granularity of the data that are processed in the functional units of an ASIP should reflect the data types in the application.

Example: The DVB-T A&T application mostly uses scalar data with a word length between 16 and 32 bit. Consequently, most of the functional units of the core use 32 bit operations. On the other hand, the FFT as well as the EVD use complex data of different bit widths requiring complex arithmetic operations (additions, subtractions, multiplications and shifts) in the functional units, which have to be designed for the maximum needed bit widths in each case.

Application	Percentage of Taken Branch Operations ($R_{branch} \cdot 100\%$)	Note on branch characteristics
DVB-T A&T	9.3%	mostly data dependent
CORDIC	10.7%	mostly needed for loops with data indep. iteration count[16]
FIR	7.1%	same as above[16]
EVD	4.1%	same as above[16]
FFT	2.1%	same as above[16]

Table 5.7: Percentage of Taken Conditional Branch Instructions (Prof. Impl.)

The bottom line of this section is that according to the above-mentioned quantitative properties of an application (instruction level parallelism, locality, control vs. data flow dominance, granularity), the task of ASIP class selection can be formalized for many different applications. This facilitates and speeds up this important design task increasing the design efficiency.

An orthogonal issue to the aspects mentioned above is verifiability of the processor hard- and software. The verification tasks and the im-

[16]The majority of these branches can be realized by zero-overhead loop control instructions which reduces the overall branch penalty to less than 2% in all these cases.

pact of architectural decisions on the verification effort is discussed in Section 5.4.

5.3 Combined ASIP HW/SW Synthesis and Profiling

The selection of the processor class in the previous subsection does not define an accurate point in the multi-dimensional ASIP design space. It rather provides a starting point for further optimizations using constraints imposed by the application and the programmability. Thus, there are still many open design issues, which need to be explored and optimized like

- organization and number of internal registers as well as memories
- behavior/coding of instructions, addressing modes etc.
- detailed pipeline organization (e.g. forwarding and bypassing) and mapping of operators to pipeline stages and functional units
- control of core operation (e.g. for startup, reset and IRQ processing)
- pipeline control policy (e.g. for branches and wait states)
- interface implementation
- optionally: coprocessor structure and implementation

Figure 5.7 shows the proposed design flow, which covers all the above-mentioned issues in order to find a feasible instruction set architecture and implementation. The main difference between the proposed design flow and previously published ASIP design approaches is that the hardware implementation in the proposed flow is in the iteration loop. This has significant consequences: On the one hand, this requires the designer to maintain several consistent descriptions of the ASIP instruction set architecture[17] (and the interfaces and coprocessors) namely for the instruction set simulator (or the system simulator, which includes

[17] This issue will be solved in Chapter 6 by a processor description language and advanced design tools which require only a *single* description of the ASIP.

interfaces and coprocessors), the software design tools and for the hardware implementation in form of a HDL. On the other hand, this methodology makes sure that the actual implementation is really able to meet the application constraints for cycle time, area and energy. In other words, this methodology enables the designer to optimize high level parameters like the cycle or instruction count of an implementation, while being able to track the effect of high level optimizations on the low level parameters cycle time, area and energy consumption. This results in less iterations and in a faster design time, provided that this design flow can be automated to a large extent.

In the following, the different subtasks of Figure 5.7 will be described with a focus on the ISA definition and on the iterative ISA optimization. This discussion also defines the requirements for the design tools of Chapter 6, which are needed to support and automate the proposed design flow of this thesis.

5.3.1 ASIP Interface Definition

Off-the-shelf processors in form of packaged chips or hard-macros are unable to adapt their interfaces to the external world. Synthesizable ASIPs, however, can easily be integrated into a system-on-a-chip, that requires a proprietary interface behavior. From a system perspective, a black box that is implemented by an ASIP can be regarded as an ordinary hardware block with typical hardware interface characteristics. Thus, the required ASIP interface mechanisms have to be negotiated between different designers or design teams. As a conventional processor is not able to handle fast streams of input data efficiently due to task switching overhead and instruction overhead in order to read the data from the input port and to store them, either specialized instructions or dedicated I/O coprocessors have to be used. For performance critical tasks with little runtime headroom for I/O operations, a DMA controller with double buffering is an option, which enables the ASIP core to focus on computations rather than on I/O activity. The detailed interface implementation can be subject to iterative refinement during the ASIP design flow in Figure 5.7.

Example: For the DVB-T A&T application a dedicated I/O processor together with instructions to support synchronization channels is needed in order to

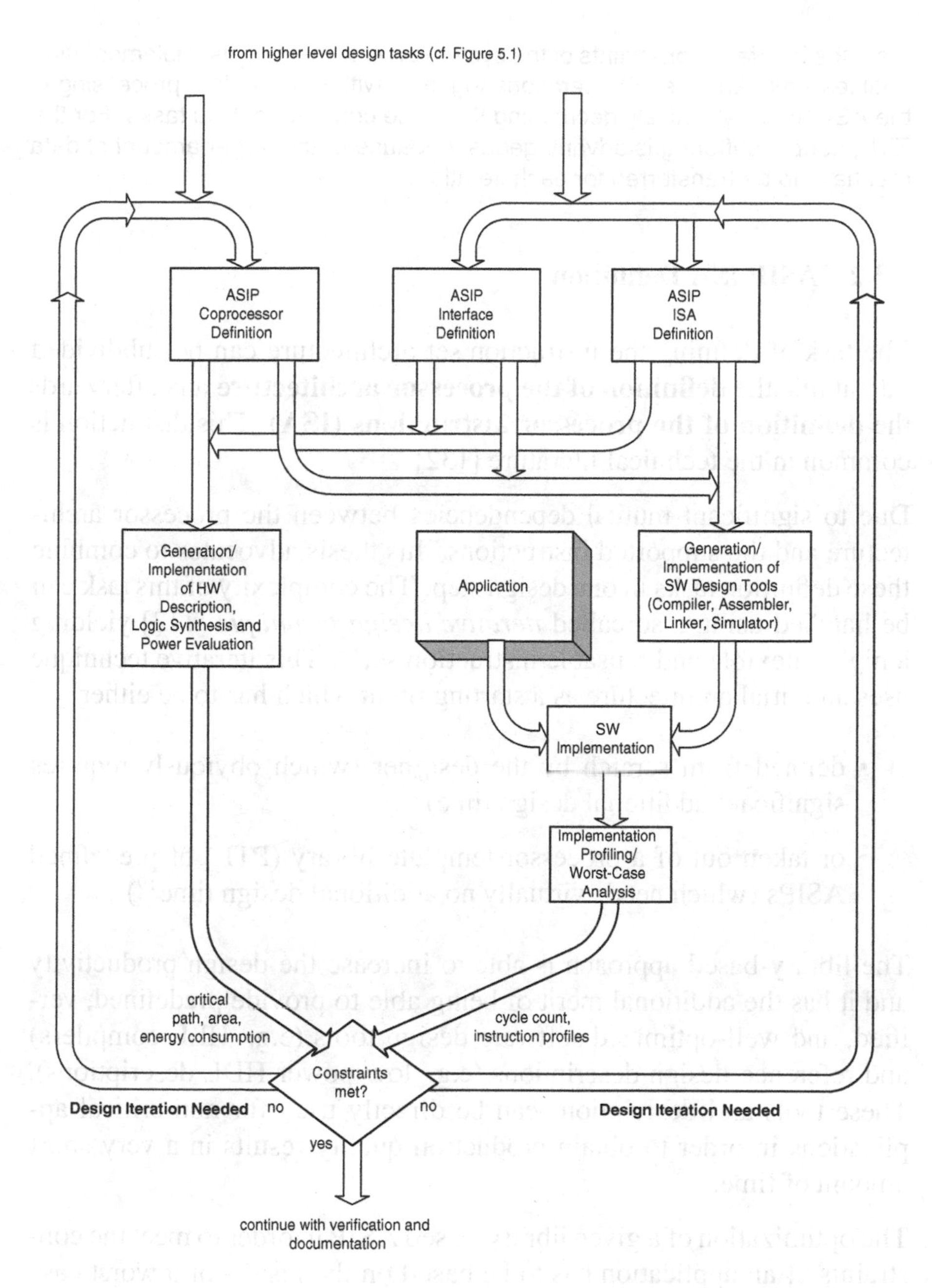

Figure 5.7: Combined ASIP HW/SW Synthesis and Profiling

meet the interface constraints of the system environment. This implementation enables simultaneous I/O operations together with normal data processing in the ASIP core, effectively decreasing the cycle count for critical tasks. For the EVD, double-buffering is advantageous, because of the larger amount of data that have to be transferred for each iteration.

5.3.2 ASIP ISA Definition

The task of defining the instruction set architecture can be subdivided into at first the **definition of the processor architecture** and afterwards the **definition of the processor instructions (ISA)**. This distinction is common in the technical literature [132].

Due to significant mutual dependencies between the processor architecture and the supported instructions, this thesis advocates to combine these definition tasks in one design step. The complexity of this task can be handled using a so-called *iterative design technique* [113] yielding a highly flexible and reusable instruction set[18]. This iterative technique uses an initial architecture as a starting point which has to be either

- defined from scratch by the designer (which obviously requires significant additional design time)
- or taken out of a processor template library (PTL) of predefined ASIPs (which needs virtually no additional design time[19])

The library-based approach is able to increase the design productivity and it has the additional merit of being able to provide predefined, verified, and well-optimized software design tools (e.g. HLL compilers) and reference design descriptions (e.g. low-power HDL descriptions). These tools and descriptions can be directly used for non-critical applications in order to obtain production quality results in a very short amount of time.

The optimization of a given library-based ASIP in order to meet the constraints of an application has to be based on the results of a worst case

[18] In contrast to so-called *constructive techniques*, which instantiate only the minimum needed amount of resources and instructions and, thus, provide a less flexible implementation.

[19] Apart from the design time which has to be spend on the design of library ASIP templates and tools by an EDA company. A similar concept is implemented in commercial logic synthesis tools like Synopsys' DesignCompiler with the DesignWare library for word-level arithmetic and logic units.

runtime analysis using the estimated low-level parameters after logic synthesis. As a function of the violated constraint(s), one or several of the techniques that have been proposed in Section 4.4 have to be applied in order to optimize the application.

5.3.3 Software Implementation and Tools

The implementation of the ASIP software requires programming tools like HLL compiler, assembler, linker as well as instruction set simulator. The implementation of these software design tools is a tedious and error-prone design task, because they have to be consistent with the ISA, coprocessor and interface definitions. In order to efficiently explore a large design space, the iteration time for the design loop in Figure 5.7 should be reasonably small. The design environment [109] that has been developed at the Institute for Integrated Signal Processing Systems, which is briefly reviewed in Chapter 6, automates the generation of these design tools.

Provided that these tools are available for a certain ASIP architecture, the software design flow is straightforward and partially comparable to commercial software design flows for off-the-shelf processors. Differences to commercial environments are due to application-specific instruction set features, accelerators and specialized interfaces, which have to be supported by the programming tools. Furthermore, typical ASIPs are used in embedded systems-on-chip that require a combination of high computational performance together with a high energy-efficiency. These design goals can only be reached, if the ASIP architecture *and* the ASIP software are jointly optimized.

Optimization of runtime of the ASIP software has to take care to fully exploit the application-specific features of the ASIP, which have been implemented to match the performance critical parts of an application. These critical software parts may have to be iteratively hand-optimized to make efficient use of gradually more specialized hardware until the performance constraints are met. The remaining part of the software that is often less performance critical should nevertheless exhibit excellent code quality in order to avoid unnecessary deterioration in overall runtime. Typical optimizations of runtime include (but are not limited to):

- avoiding redundant operations e.g. by constant propagation, common subtree removal etc.
- using dedicated instructions in order to speed up loop processing [159]
- reducing memory accesses e.g. by optimally using the available registers, register pipelining [230]
- implementing function calls by exchanging values in registers rather than using the (memory) stack
- avoiding poor schedules by considering dynamic profile data (e.g. in the case of mutually exclusive "case" selections in C code, the selection with the highest probability should be evaluated first)

Optimization of energy consumption is achieved by using optimized hardware architectures together with a software implementation that efficiently uses the hardware. In Section 4.4 the term *intrinsic energy* and the term *overhead energy* have been defined. Provided that the intrinsic energy of a task is significantly smaller than the overhead energy of a processor (which is a typical case for processors), any kind of software optimization that reduces the runtime (without increasing the overhead energy) is also lowering the energy consumption of the processor. Software performance optimization corresponds to processor energy optimization for many typical scenarios, which completely agrees to the results of Tiwari [249].

Optimization of energy consumption can be generally achieved by e.g.

- using specialized instructions that enable multiple operations per instruction
- exploiting doze/sleep modes, which disable/switch off the clock distribution/generation
- exploiting memory hierarchy e.g. by using small, low-power scratch pad memories [175]

- replacement of power greedy operations like multiplications, divisions and modulo operations for constant R-values of a power of two with more simple shift and logic operations (this strength reduction typically results only in small benefits (if any) due to the high amount of overhead power associated with each instruction)
- instruction selection based on the average energy consumed by an instruction pattern [252] [171] (with typically a very small benefit for the same reason than above)
- using a coprocessor

The above mentioned techniques can be partially integrated in the HLL compiler, but they are also useful for manual optimizations.

5.3.4 Hardware Implementation and Logic Synthesis

Estimation of low-level hardware parameters like the maximum clock speed is needed in order to guarantee the feasibility of an ASIP architecture w.r.t. the given application constraints as well as to track the effect of high level decisions during ASIP optimization. Thus, the designer has to maintain the consistency of yet another even more detailed description of the ASIP in form a HDL.

In Chapter 5.7 a tool which partially automates the generation of this ASIP HDL description is presented. This tool is able to generate a detailed hardware description of the decoder from an abstract operator-based instruction set implementation, which results in a significant speed up in design time. Further work on hardware description generation has been published in [218].

For a complete discussion of the critical factors concerning ASIP hardware implementations refer to Section 4.4.

Example: The following examples illustrate the necessity to obtain precise estimates for the hardware implementation during ASIP optimization. Logic synthesis of an initial 2 stage processor pipeline implementation for the DVB-T A&T application has indicated a maximum operating frequency that has violated the constraints. A complete redesign of the hardware implementation using a 3 stage pipeline has solved this problem. Another issue was the implementation of the general purpose register file, which initially has produced an excessive area consumption in combination with an unacceptable synthesis time. A redesign of the register file using a more structural hand-optimized HDL description (cf. Appendix E.1) has solved this problem reducing the combinational register file area by about 50%. Several times a redesign of optimized instructions using operator chaining has been necessary in order to meet the required operating frequency. There are many other examples for iterations that have been triggered by low level constraint violations. This design methodology is in analogy to best practice ASIC design flows [48], which regularly reiterate logic synthesis for estimation purposes.

5.3.5 Implementation Profiling and Worst Case Runtime Analysis

This task profiles the current HW/SW implementation considering the ASIP SW, the ASIP instruction set, the coprocessor and the interface behavior. The result of this implementation profiling is – in analogy to application profiling – the cycle count for entire tasks including I/O cycles, which has to be compared to the given constraints. After this step, the designer is aware of the critical tasks for the current implementation. Similar to the methodology used for application profiling, the critical kernels in the application have to be identified as a prerequisite for subsequent optimization.

Profiling of an implementation has to determine the worst case cycle count in order to provide an upper bound, which has to be compared to the cycle constraints of the application. This worst case cycle count can be determined either by

- simulation using stimuli, which yield this worst case condition
- (either manual or tool supported) static analysis of the code, which is especially difficult in case of many control instructions and/or computed branches

The principle of static cycle count analysis of assembler code is illustrated in Figure 5.8, which depicts the assembler implementation of a simple conditional *if*-instruction. The worst case cycle count for the implementation is given by the longest path through the assembler program which is in this case

$$C_{if\_else,wc} = max(C_1 + C_{bt} + C_{IF} + C_b,\ C_1 + C_{bf} + C_{ELSE} + C_b) \quad (5.1)$$

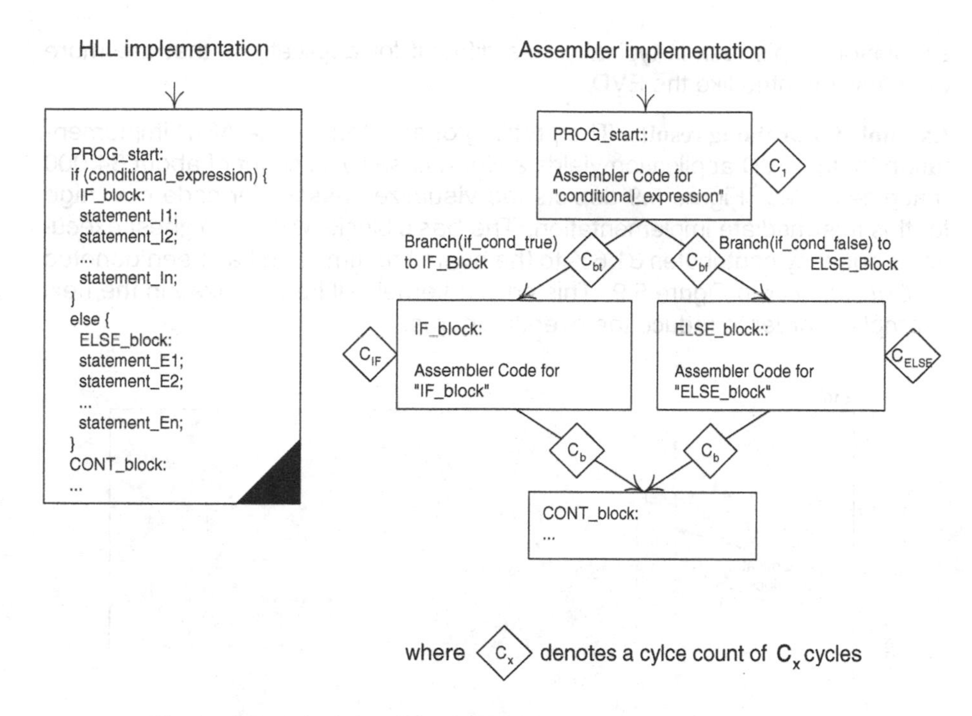

Figure 5.8: Principle of Static Worst Case Cycle Count Analysis

In the case of uncorrelated forward branches this analysis is trivial. In contrast, if (conditional) backward branches are present e.g. in order to implement HLL loop statements like *while* or *for*, this analysis is more complicated. In these cases, the designer has to determine the maximum possible number of loop iterations[20], which have to be annotated to the backward branch(es) for a worst case cycle analysis. For correlations between forward branches, the analysis in Figure 5.8 yields a possibly pessimistic upper bound in runtime. This issue is covered in more depth by Boriello [32] and Li [169].

Example for the application of the two analysis methodologies: In case of the DVB-T A&T application, simulation of typical operating scenarios has been used in the first place, in order to determine typical cycle counts. Moreover, a subsequent static cycle count analysis of this highly branch-intensive application has been manually performed to guarantee that maximum cycle constraints are met in any case. This static analysis guarantees that each path through the assembler program is covered, which is difficult to make sure with

[20]This maximum number of loop iterations can usually be derived from the HLL specification.

simulation. This task is typically less difficult for applications that are more data flow oriented like the EVD.

Example for profiling results: The profiling of an intermediate ASIP implementation for the EVD application yields a worst case cycle count of about 64 000 machine cycles. Figure 5.9 depicts the visualized assembler code coverage for this intermediate implementation. The basic blocks with the highest execution frequency contributes 31,6% to the overall runtime and has been denoted as *Critical Block* in Figure 5.9. This critical kernel will be optimized in the next example in order to reduce the overall runtime.

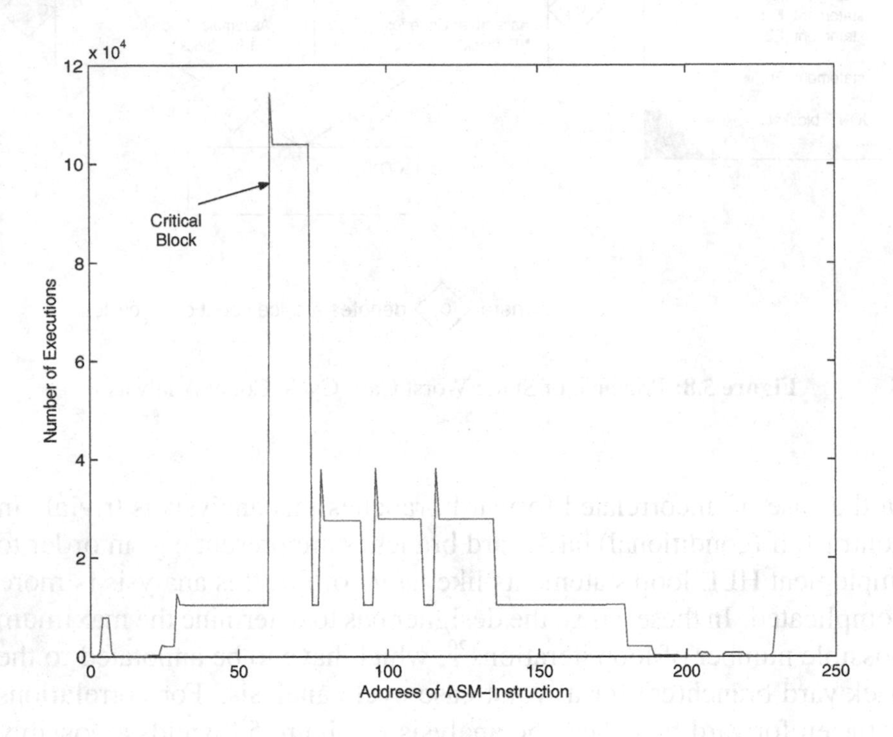

Figure 5.9: Visualized EVD Assembler Coverage (Intermediate Impl.)

5.3.6 Iterative ASIP Optimization

This optimization task primarily focuses on enhancements of the computational performance of a given ASIP architecture. However, architectural optimization can also be used in order to lower the energy consumption of an ASIP, which is demonstrated in Section 7.1.

According to Section 4.4.1 there are several options in order to increase the computational performance for critical tasks.

- chaining of operators
- parallelization of operators

which are controlled by either

- a multi-issue architecture
- or specialized instructions in a single-issue architecture each of them with the ability to control several parallel operations with just one instruction

Example (cont'd): A closer look at the critical computational loop in Figure 5.9 reveals, that the data flow graph (DFG) depicted in Figure 5.10 is executed with each loop iteration. Each of the circles in Figure 5.10 denotes an operation on complex data. The load and store operations have already been optimized by implementing a dedicated address generation, which enables the access of row- and column-indexed matrix data. The row- and column values are updated by the zero-overhead loop control logic. The assembler implementation of this kernel for the considered single-issue ASIP architecture is given in Listing 5.1. This software implementation uses 12 assembler instructions in the loop kernel corresponding to 12 machine cycles per loop iteration. The instruction at the beginning of this loop (LPINI) is used to enable zero-overhead loop control for the innermost loop by incrementing a special register after each loop iteration. Afterwards, if the end value of this loop is not yet reached, a branch back to the loop start without a delay cycle is performed.

For the DFG of Figure 5.10 optimized instructions are defined in the following in order to speed up the loop processing. In this case, these specialized instructions have to meet additional constraints of the implementation:

- only one memory port is available (which enables only one memory read or alternatively, one memory write of complex word per cycle)
- chaining of operators is impossible due to the maximum required operating frequency
- the number of operator instances and registers for temporary storage has to be minimized

This task of scheduling loop operations can be solved with software pipelining [155] e.g. using a technique like modulo-scheduling [210]. For this simple

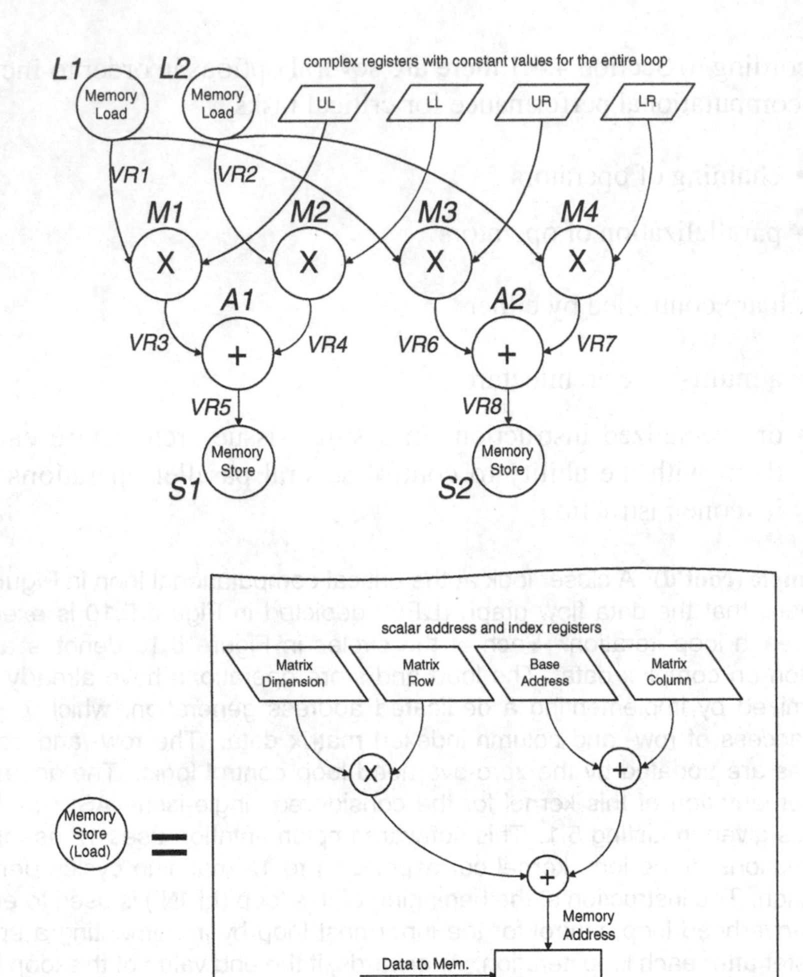

Figure 5.10: DFG of Critical Loop Kernel

example an optimum schedule and register assignment can be manually determined.

The fact that only one memory access per cycle is possible, results in a lower bound of 4 instructions for the loop body. Table 5.8 shows a possible schedule for the operations M1 to M4, A1, A2, L1, L2, S1 and S2 of Figure 5.10, which reaches this lower bound. The argument (n) refers to the processing of data

```
.define EVRLOOP_CNTR R3
  ...
  LPINI(EVRLOOP_START_LB,EVRLOOP_END_LB,EVRLOOP_CNTR,N_1,3);
EVRLOOP_START_LB:
.undef N_1 R5
.define L_FREG FR4
.define R_FREG FR5
  // load left and right columns
  FRRLD(EV_M,EVRLOOP_CNTR,ODRLOOP_CNTR,N,L_FREG);
  FRRLD(EV_M,EVRLOOP_CNTR,ODCLOOP_CNTR,N,R_FREG);

  // calculate and update left EV column
.define TMP_LEFT_COL_FREG FR6
  FMOV(L_FREG,2,TMP_LEFT_COL_FREG,2);
  // TMP_LEFT_COL_FREG = ul*EV[r][piv_row]
  FMUL(UL_FREG,TMP_LEFT_COL_FREG);
.define TMP2_FREG FR7
  FMOV(R_FREG,2,TMP2_FREG,2);
  FMUL(LL_FREG,TMP2_FREG);
  // TMP_LEFT_COL_FREG += ll*EV[r][piv_col]
  FADD(TMP2_FREG,TMP_LEFT_COL_FREG);
.undef TMP2_FREG  FR7
  FRRST(EV_M,EVRLOOP_CNTR,ODRLOOP_CNTR,N,TMP_LEFT_COL_FREG);
.undef TMP_LEFT_COL_FREG FR6

  // calculate and update right EV column
  FMUL(UR_FREG,L_FREG);
.undef R_FREG R5
.define TMP_RIGHT_COL_FREG FR5
  // TMP_RIGHT_COL_FREG  = lr*EV[r][piv_col]
  FMUL(LR_FREG,TMP_RIGHT_COL_FREG);
  // TMP_RIGHT_COL_FREG += ur*EV[r][piv_row]
  FADD(L_FREG,TMP_RIGHT_COL_FREG);
  FRRST(EV_M,EVRLOOP_CNTR,ODCLOOP_CNTR,N,TMP_RIGHT_COL_FREG);
.undef TMP_RIGHT_COL_FREG FR5
.undef L_FREG FR4
EVRLOOP_END_LB:
  ...
```

Listing 5.1: Initial Assembler Loop Implementation of Critical Kernel

in DFG iteration n, whereas the argument (n+1) means, that already data for the next DFG iteration are processed.

Cycle	Loop Instruction	Multiplier	Adder	Memory
0	PAR_INSN1	M3(n)	A1(n)	L1(n+1)
1	PAR_INSN2	M4(n)	-	S1(n)
2	PAR_INSN3	M1(n+1)	A2(n)	L2(n+1)
3	PAR_INSN4	M2(n+1)	-	S2(n)

Table 5.8: One Possible Loop Schedule for the Critical Loop Body

In analogy to the methodology used by HLL compilers [53] the edges in Figure 5.10 correspond to virtual registers which reflect the lifetime of these values in real registers. Table 5.9 depicts the lifetime of these virtual registers for the data values which are associated with the iteration number n of the DFG. It can clearly be observed from Table 5.9, that the processing for one DFG iteration is pipelined, with a latency of 8 cycles and a throughput[21] of 4 cycles.

Loop Counter	Loop Cycle	VR1	VR2	VR3	VR4	VR5	VR6	VR7	VR8
n	0								
n	1	X							
n	2	X							
n	3	X	X	X					
n+1	0	X	X	X	X				
n+1	1		X			X	X		
n+1	2						X	X	
n+1	3								X

Table 5.9: Lifetime of Virtual Registers for one DFG Iteration

These virtual registers have to be assigned to real registers. Table 5.10 shows a possible register assignment using 8 data registers.

Register Nr.	0	1	2	3	4	5	6	7
Cycle 0	UL	LL	UR	LR	*VR1	*VR2	*VR3	*VR4
Cycle 1	UL	LL	UR	LR	VR1	VR2	VR5	VR6
Cycle 2	UL	LL	UR	LR	VR1	-	VR7	VR6
Cycle 3	UL	LL	UR	LR	VR1	VR2	VR3	VR8

Table 5.10: Register Allocation for the Critical Loop Body

Note that the virtual registers VR1 to VR4 that have a '*'-prefix in Figure 5.10 are not produced in the actual but in the previous instruction loop iteration[22]. This implies that for the first loop iteration these values have to be precalculated and moved into the required real registers, which corresponds to the pipeline prologue for software pipelined VLIW-machines [7]. Also note that after the last DFG iteration two loads L1(n+1) and L2(n+1) as well as two multiplications M1(n+1) and M2(n+1) are superfluous, because they belong to the

[21] In this case, the throughput is the important aspect that has to be optimized in order to reduce the runtime. The result of 4 cycles per iteration has been obtained by neglecting the overhead due to prologue and epilogue, as well as the overhead for the loop initialization. This overhead is certainly small, because the number of iterations is large.

[22] Here, the term *instruction loop iteration* refers to one iteration of the loop that uses the optimized instructions

non-existent next iteration of the DFG. Alternatively, if this behavior of executing superfluous loads and multiplications can not be tolerated, a loop epilogue has to be implemented and the loop end count value for the optimized loop has to be decremented.

The Tables 5.8 and 5.10 implicitly define the functionality of the new optimized loop instructions PAR_INSN1 to PAR_INSN4. This new functionality is more clearly described in Table 5.11. Note that in the unoptimized implementation, the FRRST/FRRLD instructions use the loop counter in order to calculate the effective memory address according to Figure 5.10. Due to the fact that the optimized implementation uses a pipelined processing with a latency that is larger than the number of loop instructions, the FRRST instructions need to calculate the effective address using a decremented value of the loop counter[23]. This fact is considered in Table 5.11 by the notation *adr(CNT)* and *adr(CNT-1)*. Obviously, the instructions in Table 5.11 have to use many operand fields in order to implement the same functionality than the original instructions. This would lead to an unacceptable instruction coding width. In this case, these operands can be omitted using an optimized hardwired control logic due to the fact, that the reusability of these instructions is limited and that these optimized instructions do not really need the flexibility of programmable operands.

Optimized Instruction	replaces the following more simple instructions:
PAR_INSN0	FMUL FR2, FR4, FR7 || FADD FR6, FR7, FR6 || FRRLD (*adr*(CNT)), FR4
PAR_INSN1	FMUL FR3, FR5, FR6 || FRRST FR5, (*adr*(CNT-1))
PAR_INSN2	FMUL FR0, FR4, FR6 || FADD FR6, FR7, FR7 || FRRLD (*adr*(CNT)), FR5
PAR_INSN3	FMUL FR1, FR5, FR7 || FRRST FR7, (*adr*(CNT-1))

Table 5.11: Functionality of the Optimized Instructions

With these optimized instructions the enhanced software implementation of the loop is given in Listing 5.2 together with the loop prologue. This optimization has reduced the cycle count for this critical loop by 55.7% which translates in a cycle reduction of 17.6% for the overall EVD task according to Table 5.12. The moderate reduction in overall cycle count is due to Amdahl's law [9]: The critical loop which has been optimized contributes to only 31.6% and not to 100% of the total runtime. If further cycle count reduction

[23]This decremented value of the loop counter corresponds to the value of the loop counter in the previous loop iteration of the optimized implementation.

is needed, the above-described concept has to be applied to different critical blocks in the EVD task. Alternatively, the constraints that have been used for the above-mentioned example optimization have to be relaxed by increasing the hardware effort for the implementation. In any case, the feasibility of the optimized implementation has to be checked by adding the optimized instructions to the hardware description of the ASIP and by logic synthesis. Although the example optimizations have not used operator chaining in the data path, there is the risk of getting an increased critical path due to a higher number of area intensive read and write ports of the general purpose register (see Appendix E.1 for synthesis results).

```
.define EVRLOOP_CNTR R3
  ...
  // loop prologue
  MOVI(0,EVRLOOP_CNTR); // load values for 0-th iteration
  FRRLD(EV_M,EVRLOOP_CNTR,ODRLOOP_CNTR,N,FR4); // L1(0)
  FRRLD(EV_M,EVRLOOP_CNTR,ODCLOOP_CNTR,N,FR5); // L2(0)
  MOVI(1,EVRLOOP_CNTR); // start loop with iteration 1
  FMOV FR4, FR6;  // copy
  FMUL FR0, FR6;  // M1 operation
  FMOV FR5, FR7;  // copy
  FMUL FR1, FR7;  // M2 operation
  // start of loop
  LPINI(EVRLOOP_START_LB,EVRLOOP_END_LB,EVRLOOP_CNTR,N_1,3);
EVRLOOP_START_LB:
  PAR_INSN1;
  PAR_INSN2;
  PAR_INSN3;
  PAR_INSN4;
EVRLOOP_END_LB:
  // here: no epilogue
  ...
```

Listing 5.2: Enhanced Loop Implementation Using New Instructions

	Unoptimized Implementation	Optimized Implementation
Loop Cycle Count	17685	7830
Norm. Loop Cycle Count	100%	44.3%
Overall Cycle Count	63680	52475
Norm. Overall Cycle Count	100%	82.4%

Table 5.12: Cycle Count Reduction for Critical Loop

5.3.7 Definition of a tightly coupled ASIP Accelerator

The above-mentioned ISA optimization exposes limited scalability for applications that require a very large number of parallel operations. This fact is due to bottlenecks in the ASIP architecture caused by area-intensive general purpose registers or centralized data memories each of them with a limited amount of read and write ports. Furthermore, there are also applications with a significantly larger amount of operations in the critical loop bodies, which results in a large amount of optimized instructions. This in turn requires more complex decoder structures, which reduce the overall efficiency of the implementation and can lead to clock constraint violations.

For the case that an ISA optimization fails to deliver the needed computational performance or energy-efficiency, a more dedicated ASIP co-processor can be implemented. This tightly coupled ASIP accelerator can be viewed as a computationally powerful functional unit that supports either pipelined or unpipelined computations. Typically, the latency of such an accelerator is significantly larger than the latency of 1 or 2 cycles of ordinary functional units like adders and multipliers. For this reason, additional control mechanisms are needed in order to start the accelerator and to synchronize the accelerator results with the program flow of the ASIP.

Example: The EVD of an NxN hermitian matrix needs $O(N^2)$ CORDIC evaluations using the angular and the rotate mode. For a 10x10 matrix this corresponds to at least 270 angle calculations and 270 vector rotations for the desired precision. An optimized software implementation of the two CORDIC modes results in about 118000 cycles for the 10x10 EVD. By using an ASIP accelerator, this cycle count can be reduced by more than 50% to about 50000 cycles.

An important aspect for this decision is the required flexibility of a task, because a dedicated accelerator architecture is much less flexible and programmable than a pure software implementation. In case of the CORDIC subtasks for the EVD, the required flexibility is sufficiently low due to the fact that the CORDIC algorithm is a well-tested algorithm, which is not prone to late design changes and design errors.

The CORDIC algorithm (see Appendix B.1 for implementation details) uses iterative conditional additions and subtractions in order to cancel either the angle z (rotate mode) or the ordinate y (vectoring mode) of a two dimensional vector. The control data flow graph of these algorithms can be extracted and mapped to dedicated hardware. In case of the EVD, the hardware structure that is depicted in Figure 5.11 has been implemented, which supports both the vectoring and the rotate mode of the CORDIC.

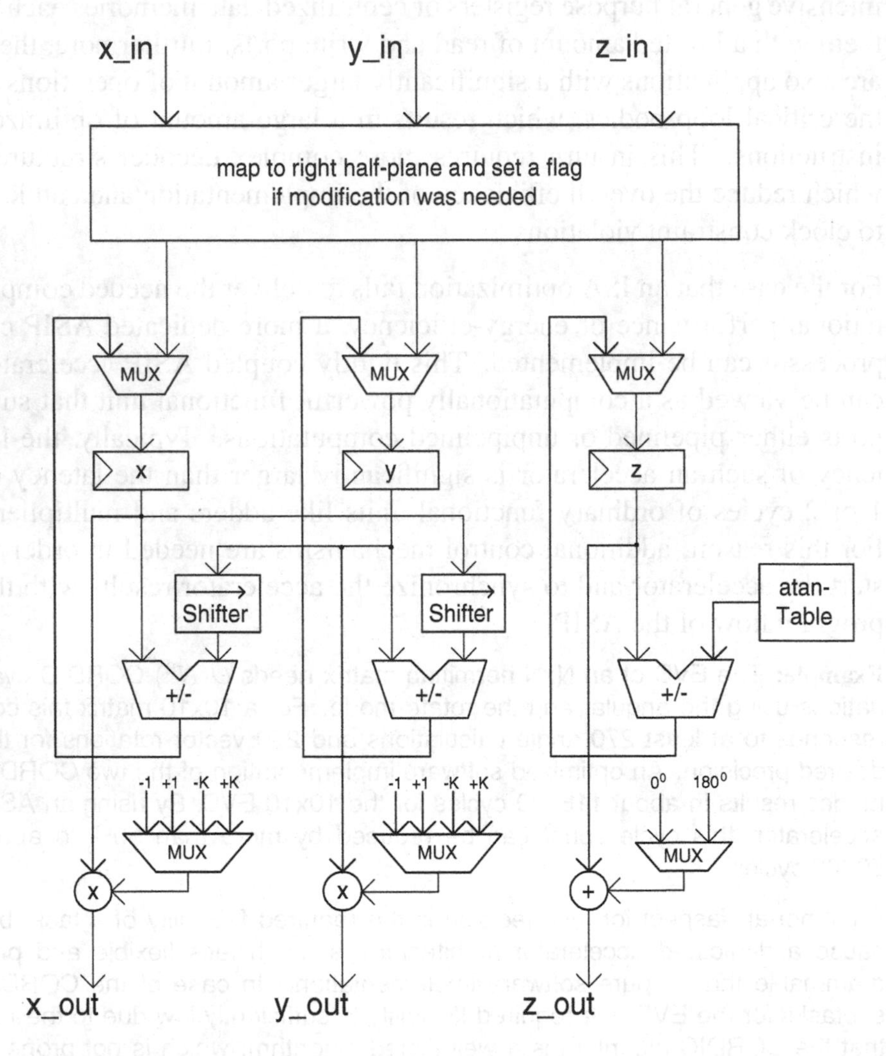

Figure 5.11: Structure of the CORDIC Accelerator

5.4 Verification

Correctness of a chip implementation prior to chip fabrication is of paramount importance due to the high prototype costs of fabrication. According to [121] the prototype costs for relative small chips between 20 sq. mm and 36 sq. mm are in the range between 600k and 900k US$. This is mostly due to the expensive production of mask sets and due to the low initial volumes for these prototypes. As a consequence, the risk of an implementation error should be minimized. This has been a matter of course for the design of dedicated hardware since many years, but this is also valid for embedded software on a chip: In the case of the DVB-T receiver described in Section 7.1 the ASIP software is stored in an on-chip ROM. In case of a software malfunction, the software in the ROM needs to be modified, which necessitates to restart the fabrication process of the chip (at slightly reduced costs) using a different mask set for the ROM information. Of course, the program information could have been stored in an internal RAM, but this would have increased the implementation power consumption and silicon area. Moreover, this would have required a more complicated bootload process for the chip, which is supposed to operate as easy to use stand-alone solution.

The term *verification* is often confused with the term *testing*. However, from a hardware perspective, testing only refers to post-silicon fabrication tests which guard against faults in the physical fabrication process. Verification rather means the process of checking, if all the design constraints (refer to Subsection 3.1.1) are met. This includes to check the behavioral equivalence of two descriptions on different levels of abstraction as well as to verify the algorithmic correctness and computational performance of an implementation.

Verification of the ASIP hard- and software w.r.t. a behavioral reference and additional time and interface constraints can be subdivided into the following three subtasks:

- Verification of the ASIP software and the ASIP instruction set simulator (ISS): Behavioral equivalence check of the ASIP SW running on an ASIP ISS for all possible input patterns. This verification task has to use a cycle-true instruction set simulator in order to determine the cycle count of the implementation, which has to meet the cycle constraints of the application[24].

After succesful completion of this task, the instance of the cycle-true ISS that has been used for this verification has also been qualified as a golden reference for the following verification of the ASIP hardware

- Verification of the ASIP hardware (processor core): Equivalence check between the behavior of the ASIP instruction set simulator (which is used as a golden reference) and the ASIP hardware implementation
- Verification of the ASIP hardware in the system environment (interfaces): The check, if the ASIP hardware interfaces comply with the specification. For this step, a model of the ASIP environment is needed, which allows to integrate the ASIP hard- and software in the system environment

Theoretically, the ASIP software can be verified together with the ASIP hardware, using a cosimulation between the ASIP simulated using an HDL simulator and the reference software implementation. However, due to the slow simulation speed of HDL simulators, this approach leads to excessive simulation runtime, which is prohibitive, if many modes of operations have to be simulated. The proposed approach separates the ASIP hardware verification and the ASIP software verification, which significantly reduces the total simulation effort. Furthermore, with this methodology it is guaranteed, that the ASIP hardware fully corresponds to the ASIP specification, which enables the designer to apply late (possibly post-silicon) design changes to the ASIP software, without having to worry about errors in the ASIP hardware.

The **ASIP software verification** process can (theoretically) be solved by exhaustive cosimulation between the behavioral reference implementation and the ASIP software running on a fast ASIP instruction set simulator. For complex applications with a large number of different operation modes and a large number of internal program states this simulation leads to prohibitive long simulation runtimes. In such a case,

[24]This verification step has to be performed in combination with a worst case cycle analysis of the final implementation according to Subsection 5.3.5.

a *divide and conquer approach* can be used, in order to verify small, independent parts of the application code[25]. An alternative would be to use a formal description as a reference implementation like in [198], but this would necessitate additional design effort (and possibly introduce additional mistakes in the formal specification), which is prohibitive to get a fast time to market. The use of a thoroughly verified, high level language compiler can significantly accelerate this verification step, because this corresponds to a *correct by construction* design approach. However, even in this case extensive functional simulations are needed in order to verify the correct behavior of the instruction set simulator, which is supposed to be used as a reference for the next verification task.

In this thesis the simulation-based verification approach is advocated, because this approach does not introduce the overhead of additional descriptions. Furthermore, the stimuli for application profiling can be reused as a basis to obtain a good application code coverage. It is also worth mentioning, that a library-based ASIP design approach significantly facilitates this design task, because the designer only needs to verify the application-specific parts of the program and the application-specific modifications of the ISS rather than the complete program[26] and the complete ISS.

Example: The verification of the software and the instruction set simulator for the DVB A&T application has been achieved using exhaustive simulation of all operating modes together with manually generated stimuli for data inputs in order to reach all relevant internal states. The instruction coverage of the software has been verified with a coverage tool, whereas the internal state coverage has been manually verified using generated histograms. In addition to these simulation scenarios, pseudo-random data inputs have been used in order to increase the level of confidence in the implementation. This verification task has required about 28% of the total design time.

The **hardware verification of the ASIP processor core** is needed to check for implementation errors in the ASIP hardware implementations w.r.t. the reference instruction set simulator. The purpose of test programs in this context is to stimulate the hardware description of the processor in order to achieve a certain coverage goal for implementation errors. Due to the high abstraction level of a RTL-based hardware

[25]In some cases, the ASIP software has to be modified in order to support this approach.

[26]Provided that the optimizing compiler for the base architecture is 100% error free.

description, this verification task is much simpler and needs much less stimuli than test vector generation for post-fabrication chip testing. On the RT-level of abstraction, the functionalities of operators like adders, shifters and multipliers are correct by definition (because they are synthesized using automatic logic synthesis[27]) and do not need to be verified exhaustively. However, the scheduling and the interconnection of all these RTL-operators and all the storage units as well as the functionality of all implicitly and explicitly described finite state machines have to be verified. Furthermore, manually designed optimized operators have to be exhaustively[28] verified.

The metric that implicitly covers parts of this error model is the code line coverage of a simulation. However, unlike the case of software verification, a full coverage of the HDL code for hardware is only a minimum requirement for verification. There are additional requirements like

a) *toggle coverage*: each binary node in the RTL-description has to switch from 0 to 1 or vice versa at least once during simulation - this metric can also extended to groups of nodes, which are required to switch to any (possible) binary combination (this metric also includes the state coverage of finite state machines)

b) for finite state machines the so called *state, transition and limited path coverage* has to verify the possible state transitions and check whether don't-care inputs can trigger wrong state transitions

c) *functional coverage* exercises a set of error-prone execution scenarios in order to check for critical events like pipeline-interlocking, data-forwarding or interrupts

Furthermore, the stimuli and the observation points for the cosimulation have to be chosen in order to maximize the effective observability-based statement coverage which was first defined in [65]. This means for instance, that test vectors that are suppressing the propagation of internal

[27]Implementation errors of the synthesis tool itself are obviously possible, but they can not be verified at this level of abstraction. In this case it is rather necessary to take advantage of formal verification between the RTL description and the synthesized netlist of a design by using commercial equivalence checkers like Synopsys' Formality [237] or CVE [33] which is an in-house tool of Infineon Tech. AG.

[28]For many manually designed operators, (nearly) exhaustive verification is in fact possible, due to the low complexity of these operators.

incorrect values to the observation points should be avoided. A good overview of verification issues is given in [241].

The approach that is advocated in this thesis for the RTL design verification is based on the work that the author has published in [87]. Starting from an instruction set architecture definition and a set of user-defined rules, constraints, and test biases, a test case generator (TCG) is used to generate the test programs that satisfy the above-mentioned constraints. The selected approach enables the support of significantly different architectures[29], because the user has direct control over the test case generation process. A detailed description of the TCG tool is given in Subsection 6.3.2.

Example: For the DVB-T A&T application automatic test program generation has been used in order to verify the behavior of the HDL description for boundary conditions. Furthermore, functional cosimulation using a subset of the already available stimuli for software verification has been performed in order to simulate the typical behavior of the implementation. These test programs and test stimuli have been added to a regression test suite to verify the functionality of the implementation after design changes.
It has to be emphasized that the DVB-T A&T implementation has been designed in order to ease verification. This has been achieved by a largely orthogonal base implementation complemented by application-specific functional units. The orthogonal base architecture and the application-specific functional units have been verified separately. Unorthogonal features like multi-cycle, multi-word instructions, and complicated internal state machines have been avoided. This significantly eases the debugging process for the hardware implementation during cosimulation with the bit- and cycle-true instruction set simulator. This hardware verification task has required about 11% of the total design time.

The last verification task, namely the **verification of the ASIP hardware interfaces** is needed to check, if the ASIP interfaces comply with the constraints of the system environment. The different verification tasks concerning these interfaces are

- the interface protocol constraints
- the low level timing constraints of the final synthesized, placed and routed design
- correct interconnections

[29]This methodology has been succesfully applied to a TMS320C25 DSP clone in [87] (accumulator based architecture) and to the DVB-T A&T processor of the case study in Section 7.1 (load/store architecture).

The low level timing constraints have to be checked after synthesis of the complete system and/or after place and route. The interconnections between the ASIP and the system environment as well as the correct implementation of the interface protocols require a simulation of the ASIP hardware description (running the ASIP application software) in a model of the system environment. This model of the system can either be a monolithic RTL-based hardware description or a heterogeneous set of behavioral or RTL-based hardware blocks, which can be simulated within a commercial system simulation environment like e.g. Synopsys' CoCentric System Studio [238] or Cadence's VCC [39].

Example: For the simulation of the complete DVB-T system including all digital parts of the DVB-T receiver, RTL-VHDL simulation on a high end workstation with multiple parallel CPUs has been used. This verification task has required about 6% of the total design time.
For even more complex systems in the future, however, this methodology might become difficult due to excessive simulation runtime. Models that use a higher level of abstraction and enable faster simulation might be a solution for this issue.

5.5 Concluding Remarks

This chapter has introduced the proposed ASIP design flow of this thesis. In contrast to previous ASIP design approaches, the ASIP hard- and software is in the main design iteration loop of the proposed design methodology. This implies that many of the tedious ASIP design tasks have to be automated to a large extent in order to obtain a short time-to-market. The next chapter briefly describes the LISA tool suite, which is able to meet this requirement. A special focus of the next chapter are new concepts for hardware generation and verification, that have been triggered by this thesis.

Chapter 6

The ASIP Design Environment

This section starts by giving an overview of the LISA[1] processor description language and the tools that can be generated by the LISA design environment[2]. The focus of this chapter are the concepts and tools for ASIP-specific extensions to this former design environment that have been developed in this thesis. The features of the latest LISA tool suite (status of october, 2003) are summarized in Appendix A.

6.1 The LISA Language

The LISA language is based on two different language constructs, namely

- *resources*, which declare storage units like data and address registers, pipeline registers etc. as well as the memory organization
- *operations*, which define the instruction syntax, the instruction coding and the state transitions that are performed by instruction execution

The *RESOURCE* section is a straightforward description of the processor resources, using a syntax which resembles the definitions of variables in the high level language C. Listing 6.1 shows an excerpt of a RESOURCE section for the scalar part of the ICORE-II architecture described in Section 7.2.

Listing 6.1 demonstrates the usage of data types for fixed point arithmetic with arbitrary bit width (the *bit* data type), which is one important feature of LISA for ASIP design. The bit width is one key parameter

[1]Language for Instruction Set Architecture Description [286]

[2]This summary refers to the LISA tools as available, when this thesis was started. In the meantime, many enhancements proposed by this thesis have been implemented in the production version of the LISA tools.

```
RESOURCE // EVD-PAST-Processor
{
  MEMORY_MAP
  {
    0x0000 -> 0x0200, BYTES(3) : prog_mem[0x0000..NUM_PROGMEM_WORDS-1],
                      BYTES(3);
    0x0000 -> 0x0100, BYTES(4) : data_mem_r[0x0000..NUM_DATAMEM_WORDS-1],
                                 BYTES(4);
    0x0000 -> 0x0100, BYTES(4) : data_mem_i[0x0000..NUM_DATAMEM_WORDS-1],
                                 BYTES(4);
  }

  PROGRAM_MEMORY long prog_mem[0x0000..NUM_PROGMEM_WORDS-1];
  DATA_MEMORY    signed bit[MEM_WL] data_mem_r[0x0000..NUM_DATAMEM_WORDS
                                               -1];
  DATA_MEMORY    signed bit[MEM_WL] data_mem_i[0x0000..NUM_DATAMEM_WORDS
                                               -1];

  PROGRAM_COUNTER unsigned int PC;         // normal PC
  REGISTER        unsigned int BPC;        // PC for Branch Processing
  REGISTER        unsigned int OPC;
  REGISTER                 bool BPC_valid;

  REGISTER signed bit[DP_WL] FR_r[0..NUM_FREGISTERS-1];
  REGISTER signed bit[DP_WL] FR_i[0..NUM_FREGISTERS-1];
  REGISTER signed bit[ld_NUM_MEM_WORDS] R[0..NUM_IREGISTERS-1];

  PIPELINE pipe = { FE; DE; EX };
  PIPELINE_REGISTER IN pipe {
    long instr1, instr2, instr3, instr4; /* 24 bit words */
    int  pc;
  };

  long cycle, instruction_counter;

  /* zero overhead loop support */
  int ZOLP_active[0..NUM_NESTED_ZOLP-1];
  int ZOLP_start_addr[0..NUM_NESTED_ZOLP-1];
  int ZOLP_end_addr[0..NUM_NESTED_ZOLP-1];
  int ZOLP_R_end_value[0..NUM_NESTED_ZOLP-1];
  int ZOLP_increment_flag[0..NUM_NESTED_ZOLP-1];

  ...
}
```

Listing 6.1: Example RESOURCE Section

that has to be tailored to an application in order to reduce the hardware overhead. Furthermore, the RESOURCE section supports constants, which can be included from a standard C header file in order to parameterize the implementation. This helps the designer to keep the description consistent, which has been proved to be useful in hardware description languages like VHDL (the *GENERIC* parameters) and Verilog (the *parameter* values). In case of multiple operations per pipeline stage the keyword *REGISTER* used together with a compiler flag enables cycle accurate behavior of clocked resources. This feature is needed in order to obtain an unambiguous result regardless of the order of executed operations[3]

The *OPERATION* section in LISA is used to describe the state transitions of the resources by specifying the properties of the processor instructions. For this purpose, the *OPERATION* section has several subsections:

- a *DECLARATION* section, which declares instances or groups in order to construct a complete tree of operations for one processor instruction
- a *CODING* section, which defines the coding for the actual operation resulting in a coding tree for the complete instruction set of the processor
- a *SYNTAX* section, which defines the assembler syntax of the actual operation resulting in the complete syntax definition for the instruction set
- a *BEHAVIOR* section, which defines the state transitions performed by an operation
- an *ACTIVATION* section, which triggers the behavior of other operations
- an *EXPRESSION* section, which returns values to parent[4] operations
- a *SEMANTIC* section, which will be used for future compiler support in order to provide additional information to a future compiler generator

[3]Without this flag and under certain conditions, LISA is prone to race conditions, which result in an incorrect result of clocked resources. A simple example is a flag that is modified by one operation in a pipe stage and that is read by another operation in the same pipe stage. Without the clocked behavior, the updated (and wrong) value is read (if the modified operation has been executed first), whereas in case of clocked behavior the old registered (correct) value is read in any case. This feature is analogous to race conditions in Verilog, which have to be avoided by the designer.

[4]The parent of an operation refers to the tree which represents the hierarchy of operations. This tree can be seen as a decomposition of an instruction into smaller parts e.g. register fields associated to register read operations, memory fields associated to memory accesses etc.

Listing 6.2 shows an extract for the description of a coding subtree, which is used to describe the branch instructions of the processor in Section 7.2. In this listing the operation *insn_branch* is the parent of the operations *insn_branch_cond* and *insn_branch_uncond*.

```
OPERATION insn_branch IN pipe.EX
{
  DECLARE
    { GROUP insn = { insn_branch_cond || insn_branch_uncond };
      GROUP adr = { address_op }; }
  CODING { 0b00 0bx[1] insn adr }
  SYNTAX { insn adr ")" }
  BEHAVIOR
    {
      adr();   // perform address calculation before instr. execution!
      insn();  // (addresses are temporarily stored in global variables)
    }
}
OPERATION insn_branch_cond IN pipe.EX
{
  DECLARE
    { GROUP insn = { FBZ || FBARGE }; }
      // cond. branch
      // if register zero (FBZ)
      // if abs(real(register1)>=register2
  CODING { 0b1 0bx[3] insn }
  SYNTAX { insn }
  BEHAVIOR { insn(); }
}

OPERATION insn_branch_uncond IN pipe.EX
{
  DECLARE
  { GROUP insn = { B || LPINI }; } // uncond. branch/init. zero ovhd. loop
  CODING { 0b0 0bx[3] insn }
  SYNTAX { insn }
  BEHAVIOR { insn(); }
}

OPERATION FBZ IN pipe.EX
{
  DECLARE
  { INSTANCE freg; }
  CODING { 0b0 freg 0bx[3] }
  SYNTAX { "FBZ" "(" freg "," }
  BEHAVIOR { if ((FR_r[freg].ExtractToLong(0,LONGBITS)==0) &&
                 (FR_r[freg].ExtractToLong(0,LONGBITS)==0)) {
               BPC_valid = 1; BPC = address_tmp_var;
               PIPELINE_REGISTER(pipe, FE/DE).flush();
               PIPELINE_REGISTER(pipe, DE/EX).flush();}
           }
}
...
```

Listing 6.2: Example Coding Tree Description for Branch Instructions

In Listing 6.3 the coding root for the same processor is described, which supports 24/48 bit instruction word widths. The root of the coding tree is defined by the keywords *CODING AT*. The *SWITCH* statements decides as a function of one bit in the coding, whether a one or a two word instruction is selected (*insn_1word* or *insn_2word*).

```
OPERATION decode IN pipe.DE
{ DECLARE { ENUM InsnType = { type_1word, type_2word};
            GROUP Insn_grp_1word = { insn_1word };
            GROUP Insn_grp_2word = { insn_2word }; }
  SWITCH (InsnType)
  {
    CASE type_1word:
    {
      CODING AT (OPC) {
        PIPELINE_REGISTER(pipe, FE/DE).instr1 == Insn_grp_1word}
      SYNTAX {  Insn_grp_1word }
      ACTIVATION {  Insn_grp_1word }
    }
    CASE type_2word:
    {
      CODING AT (OPC) {
        (PIPELINE_REGISTER(pipe, FE/DE).instr1 == Insn_grp_2word=[24..47])
     && (PIPELINE_REGISTER(pipe, FE/DE).instr2 == Insn_grp_2word=[0..23])}
      SYNTAX {  Insn_grp_2word }
      ACTIVATION {  Insn_grp_2word }
    }
  }
}

OPERATION insn_1word IN pipe.DE
{
  DECLARE
    { GROUP insn = { insn_branch || insn_one_cycle ||
                     insn_multi_cycle ...}; }
  CODING { 0b0 insn }
  ...
}

OPERATION insn_2word IN pipe.DE
{
  DECLARE
    { GROUP insn = { FSI_VEC || FMMMUL_VEC || FMADD_VEC ||
              FMSUB_VEC ... };}
  CODING { 0b1 0b00 insn }
  SYNTAX { insn }
  BEHAVIOR { PC = PC + 1; OPC = OPC +1;
    // insn(); /* replaces the following activation in this case */
            }
  ACTIVATION { insn }
}
```

Listing 6.3: Example Description of Coding Root

The LISA language has the ability to describe pipelined architectures. A specific operation can be assigned to a certain pipeline stage using the keyword *IN* together with a declaration of the stage names in the RESOURCE section (see Listing 6.1). The operations in Listing 6.2 are executed in the pipeline stage EX, whereas the operations in Listing 6.3 are executed in stage DE.

The assembler syntax for one instruction is described by the ensemble of all operations that describe the behavior and coding of this instruction. As an example, Listing 6.4 describes the syntax of the instruction FADD: A legal instance of this instruction is e.g. FADD (FR1, FR7).

```
OPERATION insn_2freg IN pipe.EX
{
  DECLARE
  {
    GROUP fregs1, fregd1 = { freg };
    GROUP insn = { FADD  || FSUB || FMUL };
  }
  CODING { 0b001 0bx[4] insn fregs1 fregd1 }
  SYNTAX { insn "(" fregs1 "," fregd1 ")"  }
  ACTIVATION { insn }
}

OPERATION freg IN pipe.EX
{
  DECLARE { LABEL index; }
  CODING { index=0bx[3] }
  SYNTAX { "FR" ~index=#U }
  EXPRESSION { index }
}

OPERATION FADD IN pipe.EX
{
  DECLARE
  {
    REFERENCE fregs1, fregd1;
  }
  CODING { 0b000 0bx[4] }
  SYNTAX { "FADD" }
  BEHAVIOR { FR_r[fregd1] = FR_r[fregd1] + FR_r[fregs1];
             FR_i[fregd1] = FR_i[fregd1] + FR_i[fregs1]; }
}
...
```

Listing 6.4: Example Syntax Description of FADD Instruction

Listing 6.4 is also an example for the usage of an *EXPRESSION* section, which returns the numeric value of the 3 bit register field *freg* to the operation *insn2_freg* in this case. These values are not directly needed in operation *insn2_freg*, but rather referenced (keyword *REFERENCE*) and used by the operation *FADD* in order to perform the actual complex addition.

Listing 6.3 depicts an example for the *ACTIVATION* section, which is used in order to trigger the execution of further instructions that are declared after the *GROUP* keyword. In this case the ACTIVATION section in Listing 6.3 can be replaced by an explicit call to *insn()* in the *BEHAVIOR* section in this example. The order of execution of *ACTIVATION* and *BEHAVIOR* section depends on a LISA compiler switch. For the current example, this switch is set to execute the *BEHAVIOR* section first.

6.2 The LISA Design Environment

Development tools for software and hardware are of paramount importance for ASIP designs in order to efficiently profile the applications and architectures and to obtain error-free implementations. The application and architecture profiling methodology in Chapter 5 requires a retargetable compiler as well as a simulator with profiling capabilities. Furthermore, hardware generation using a high level architecture description is beneficial to reduce the design time.

The LISA ASIP design environment uses a single LISA description in order to generate the following software design tools: assembler, linker and simulator with API as well as debugger, debugger GUI, profiler and cosimulation interfaces. Figure 6.1 provides an overview of this design environment: The LISA processor compiler[5] is the heart of this environment and generates the design tools automatically according to Figure 6.1.

The generated tools support a large part of the software development process for ASIPs, currently with the exception of a HLL compiler, which is subject to ongoing research.

[5]The name *LISA Processor Compiler* does not refer to a high level language compiler, which generates assembly code for a processor. The LISA processor compiler is rather responsible to generate the abovementioned software design tools.

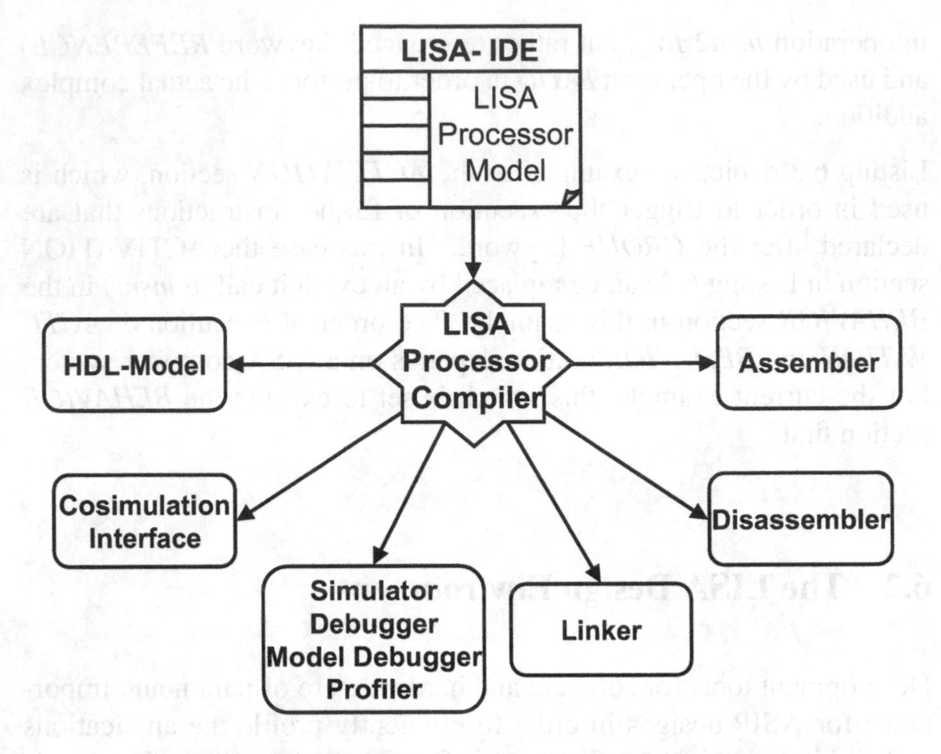

Figure 6.1: The LISA Processor Design Environment

A significant part of the overall design time according to Appendix F is needed for the hardware description and verification of an ASIP. Recently[6], the hardware generation task has been fully automated by the LISA HDL generator, which has been developed at the Institute for Integrated Signal Processing Systems [218].

This HDL generator uses the pipeline description in the LISA RESOURCE section to automatically generate the ASIP pipeline registers as well as a coarse structure of the ASIP. Furthermore, the decoder is generated using the information of the CODING sections. Empty wrappers for the functional units can be automatically obtained, whereas the generation of the internal structure of these functional units is currently being developed. In order to achieve this additional functionality, the

[6]When I wrote the first version of this thesis, this LISA task was still only partially automated. Currently, a large part of the enhancements for HDL generation described in the following sections are already fully functional in the LISA production tools.

LISA keyword UNIT has been introduced in order to bind operations to functional units.

The LISA processor design methodology as available, when this thesis has been started, had two considerable disadvantages:

- the designer had to provide full and consistent CODING information even during design exploration, which often required changes in the complete coding tree, especially if additional instructions were inserted
- tedious verification of the complete ASIP description, consisting of automatically generated and hand-written parts

Enhancements to this design methodology are presented in the following section, which remove these disadvantages and enable to obtain optimum results within a significantly reduced design time.

6.3 Extensions to the LISA Design Environment

In this section, two extensions to the LISA design environment are described that have been developed in this thesis: An instruction encoding and decoder generator, as well as a semi-automatic test pattern generator. Apart from the speed up in design time for both approaches, the instruction encoding generator automatically produces instruction encodings with an optimum coding density to reduce the instruction memory size. Furthermore, significant energy savings for a given program will be demonstrated by using encoding optimization that takes into account the profiling information of the program.

6.3.1 Instruction Encoding and Decoder Generation

Initially[7], the LISA language required the user to specify the detailed coding of each instruction right from the beginning of a design space

[7]The current version of the LISA tools includes a fully functional automatic coding capability as suggested by this thesis.

exploration. This had the consequence of tedious manual modifications in complete coding subtrees, if new instructions were inserted.

Automatic coding generation is useful in order to speed up this process: The user only has to specify the instruction operands, whereas the instruction opcode field can be omitted. The detailed instruction coding is generated automatically exploiting the information of used operands that is provided by the *DECLARE* section. After this process is finished, the LISA compiler can generate the software programming tools as usual. Furthermore, the hardware description of the decoder can easily be generated. For this thesis, an experimental EDA tool for this task has been implemented, which is referred to as ICON[8] in the following discussion.

The percentage of the instruction memory power of an ASIP can be significant according to the results of the case study in Appendix F. For applications with larger instruction memories (or instruction caches implemented by RAMs) this percentage is obviously even higher. The ASIP instruction coding directly affects the size and the energy consumption of the instruction memory. For this reason, an additional automatic optimization step to instruction coding has been developed.

The instruction coding affects the silicon area for the program memory as well as the energy consumption for the following reasons:

- the power consumption and the area of the instruction memory are approximately proportional to the instruction width, provided that the access schemes and the toggle activity are constant
- a large part of the energy consumed in the instruction memory depends on the toggle activity of the internal bit lines (refer to Figure 6.5), which represent large capacitances
- in case of external instruction memory, the instruction bus is even more heavily loaded by pad and external capacitances leading to a considerable power contribution

As a consequence, ICON has two different optimization tasks:

- minimization of the instruction width

[8] Instruction COding geNerator

- minimization of either the
  - internal memory toggle activity or the
  - toggle activity on the instruction bus

The following definitions of important terms w. r. t. instruction coding are used in this discussion:

- the term *instruction* (or *instruction instance*) represents the idea of one specific ASIP instruction with an associated behavior e.g. MOVI #3, R5 (move the value 3 to register 5)
- *instruction operand* refers to the operands of an instruction like e.g. #3 or R5 in the above example
- the term *instruction word* or *instruction code word* refers to the coded representation of one instruction e.g. "000111001010..."
- the term *instruction type* is the generic term for the set of possible instructions with the same behavior and the same operand types like e.g. addition of an immediate to a register value (represented by thc mncmonic *MOVI*)
- the term *operation code* or *opcode* refers to the part of the instruction code word, which determines the instruction type
- the *instruction format* determines the position of the opcode and the operand fields within the instruction word for each instruction type

For the sake of simplicity and due to the fact that many ASIPs use simple fetch and decoding units, only constant length instruction words are covered in the following.

6.3.1.1 Minimization of the instruction width

This task can be performed with or without a reduction in flexibility of the instruction set. The lower bound of the instruction word width is given by the width of the binary word which is needed to enumerate all the different instruction instances that are used in a given program. If we count $N_{used\_instructions}$ different instructions in a given program this results in the minimum width of

$$W_{instr,min} = \lceil \mathrm{ld}(N_{used\_instructions}) \rceil \quad (6.1)$$

This minimum instruction word width has the demerit of massively reducing the flexibility and reusability of the instruction set, because program changes requiring new instructions with different operand values are impossible. Furthermore, this methodology necessitates a significant effort for decoding, which is prohibitive for logic synthesis in case of many different supported instructions.

On the other hand, a reasonable upper bound[9] for the instruction word width can be obtained by including all the different operand fields each of width $W_{operand,i}$ and the opcode field of width W_{opcode} for each instruction type into the instruction code word and performing the maximum operation, which results in a width of

$$\begin{aligned} W_{instr,max} &= \max_{\forall opcode\ types} \left(W_{opcode} + \sum_i W_{operand,i}\right) \\ &= \max_{\forall opcode\ types} \left(W_{opcode} + W_{operand\ fields}\right) \end{aligned} \quad (6.2)$$

The opcodes that are assigned to the $N_{instr\ types}$ different instruction types can use e.g.

- one-hot encoding [46] which yields (the upper bound of the opcode width of)

$$W_{opcode,one\_hot} = N_{instr\ types} \quad (6.3)$$

- constant width encoding by simple enumeration resulting in

$$W_{opcode,enum} = \lceil \mathrm{ld}(N_{instr\ types}) \rceil \quad (6.4)$$

- prefix encoding using code words of different widths (similar to the concept of a Huffman-code [119], resulting in the lower bound for the opcode width as demonstrated later on)

[9] More coding bits would represent a waste, but are obviously possible.

The one-hot encoding approach is only useful in order to get an extremely simple decoder in hardware, which is irrelevant to the considered decoder complexities in the context of ASIPs with a reasonable amount of different instruction types (typ. < 100 instruction types). For practical implementations either the prefix or the constant width encoding approach is favorable in order to minimize the instruction word width.

After the ASIP designer has determined the useful instruction operands[10], the tool ICON optimizes the instruction opcode assignments. For the prefix code, ICON uses an algorithm similar to the one for constructing a Huffman code (cf. Cormen [59], pp. 339-341). Instead of the symbol probability of the Huffman approach, the total width of the operand fields $W_{operand\,fields,j}$ of each instruction j is used to build the Huffman tree. Consequently, the annotation of the new node has to be performed using the modified update function for the so-called merge operation

$$w(z) = max(w(x), w(y)) + 1 \tag{6.5}$$

instead of the simple addition of probabilities in the original Huffman tree construction algorithm. This new update function reflects the fact, that the new subtree has the coding width of the maximum of the subtrees incremented by one for an additional decision bit between the right and the left subtree. Listing 6.5 depicts the resulting algorithm, which constructs an optimum opcode assignment. The optimality follows from the optimality of the greedy algorithm for the original Huffman coding problem [59].

For instruction sets with differences between the sum of the used operand field widths $W_{operand\,fields}$ for different instruction types, this methodology typically yields a shorter overall coding width W_{instr} than the constant opcode width approach. Otherwise, the width is equal to the constant opcode width encoding approach.

Example: Consider the instructions, which are depicted in Figure 6.2. They have the operand field widths $W_{operand\,fields,i=0\cdots7} = (6, 19, 13, 3, 9, 0, 9, 9)$ (the empty fields represent unassigned bits in the instruction word).

---

[10] Omissions of operand-fields for certain instructions are possible, but reduce the flexibility and orthogonality of the instruction set. This critical decision should be left to the ASIP designer.

```
/* C is a set of n different instruction types with
   associated operand widths                            */

MOD_HUFFMAN (C)
1  n = |C|
2  Q = C                                  // Q is a prority queue
3  FOR i=1 TO n-1
4     DO z = ALLOCATE_NODE                // -create new node instance
5        x = left(z)  = EXTRACT_MIN(Q)    // -extract the two min. nodes
6        y = right(z) = EXTRACT_MIN(Q)    //  operand widths from Q
7        w(z)  = max(w(x),w(y))+1         // -new update function
8        INSERT(Q,z)                      // -insert new subtree in Q
9 RETURN EXTRACT_MIN(Q)                   // -return root of Huffman tree
```

Listing 6.5: Algorithm for Optimum Opcode Coding Tree Construction

Instruction	Bits 19–0
0: MOV Rs, Rd	Rs (bits 7–4), Rd (bits 3–0)
1: MOVI #imm, Rd	imm (bits 18–4), Rd (bits 3–0)
2: R (ARn,$offset), Rd	offset (bits 12–6), ARn (bits 5–4), Rd (bits 3–0)
3: CLR Rd	Rd (bits 3–0)
4: MUL Rs1, Rs2, Rd	Rs1 (bits 11–8), Rs2 (bits 7–4), Rd (bits 3–0)
5: SLEEP	
6: ADD Rs1, Rs2, Rd	Rs1 (bits 11–8), Rs2 (bits 7–4), Rd (bits 3–0)
7: SUB Rs1, Rs2, Rd	Rs1 (bits 11–8), Rs2 (bits 7–4), Rd (bits 3–0)

Figure 6.2: Operand Widths of Example Instruction Set

A constant width encoding approach requires 3 bits in the opcode field to code the 8 different instruction types, which results in an overall instruction width of 19+3=22 bits (19 bits are needed for the operands of instruction 1). The proposed algorithm for prefix encoding requires only 20 bits according to Figure 6.3. The final instruction coding for this prefix encoding approach is depicted in Figure 6.4. Here, it is obvious that the instruction 1 determines the overall instruction width due to the long intermediate field.

For the real-world instruction set in Appendix C this coding assignment optimization also yields a 10% reduction in coding width. It has to be emphasized, that this optimization does not impair the flexibility and reusability of the instruction set.

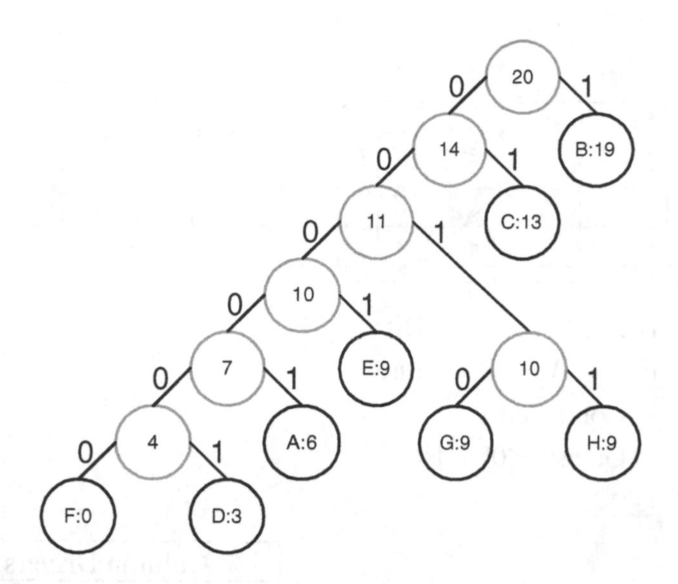

Figure 6.3: Coding Tree for Example

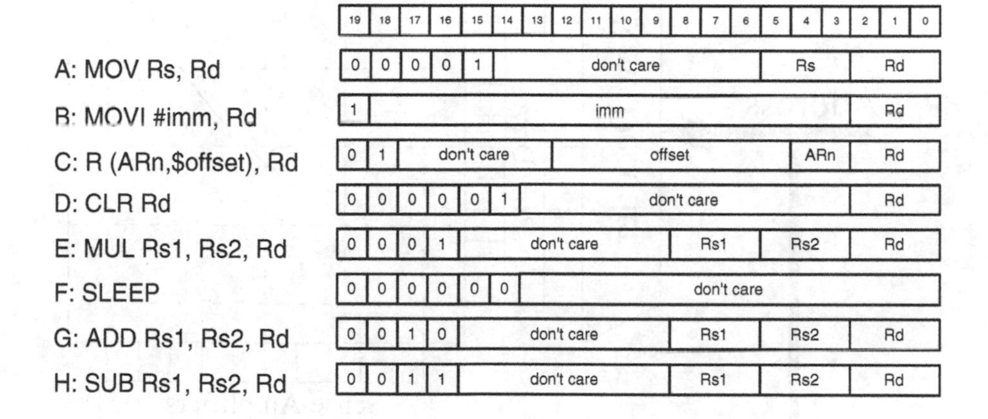

Figure 6.4: Final Instruction Coding for Example

6.3.1.2 Minimization of the Toggle Activity

The first toggle activity optimization that is described here is the optimization of the toggle activity for **on-chip instruction memories**. In Figure 6.5 the internal structure of a read-only memory is depicted[11]

[11]The depicted ROM uses NMOS-bit cells. RAMs use a comparable structure with different memory cells.

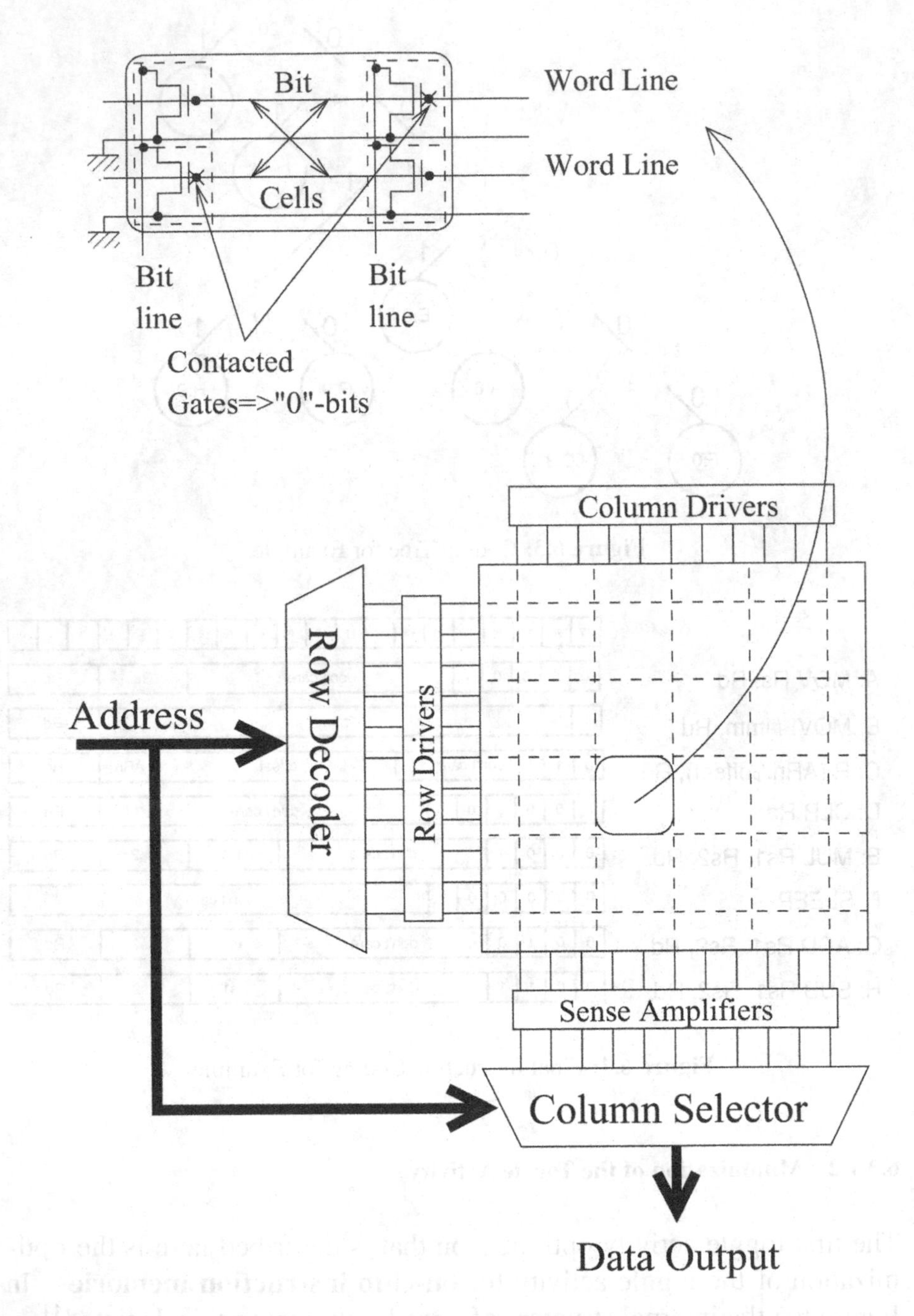

Figure 6.5: Internal Structure of a ROM

Typical NMOS-ROMs use a two-phase access scheme, which starts by precharging the bit lines to logic 1 in the first phase. The access to a row of bit cells is performed in the second phase by the row decoder which asserts one word line. If a specific bit cell contains a logic 0, the associated bit line is decharged. Otherwise, no decharging activity occurs and the bit line remains in the charged state. Figure 6.6 shows a model of the ROM with the relevant internal capacitances using ideal switches for the bit cells. This ROM model has been used for the power evaluations in this thesis. According to the case study in [47] 70% of the total energy consumption of an SRAM is required for the bit lines, the associated sense amplifiers and the bit cells themselves[12]. In the following, we assume 30% to 60% for the percentage of power consumed by the bit line toggle activity of a ROM.

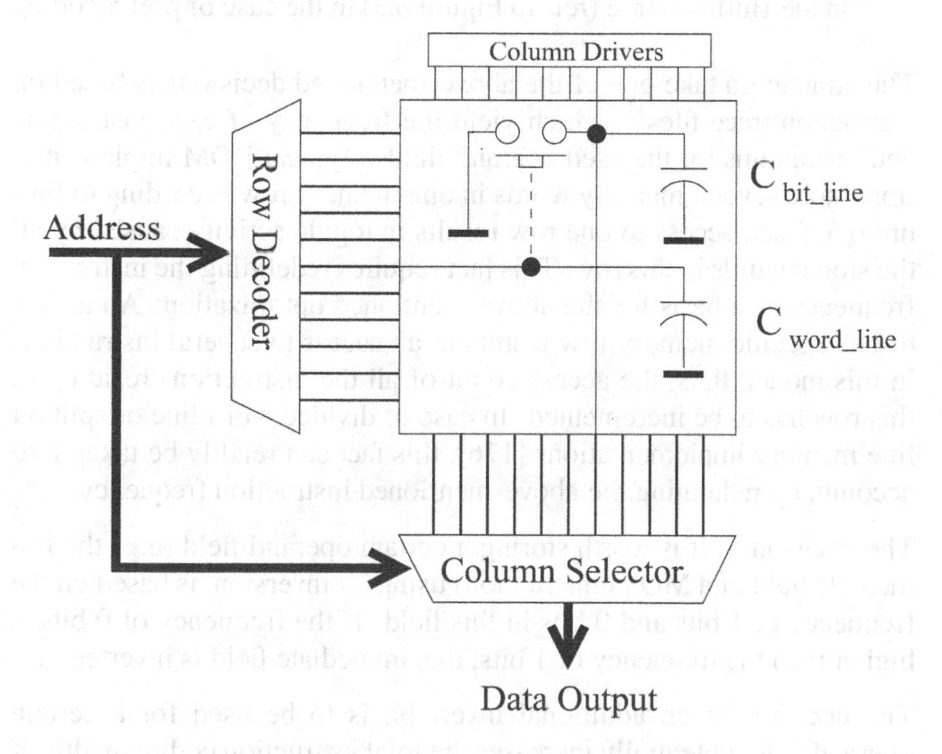

Figure 6.6: ROM Model with Capacitances

[12]This figure strongly depends on the organization of the memory: shorter bit lines can be traded-off for longer word lines. Furthermore, divided bit and word lines or multi-block partitioning can be used [257].

The proposed algorithm to find an optimum instruction encoding takes advantage of the fact that from an energy perspective the 0 bits in the memory are more expensive than the 1 bits. A straightforward conclusion is to assign 1s to each don't care bit in each instruction. Further degrees of freedom for this code assignment are:

- use of non-inverted or inverted operand fields as a function of the specific instruction type
- use of a redundant operand representation with one additional *invert bit*, which indicates, that the stored value has been bit inverted
- assignment of maximum weight first opcode codings, in case of constant width encoding, or swapping of the binary decision bits in the Huffman tree (ref. to Figure 6.3) in the case of prefix coding

The strategy to take one of the above-mentioned decisions is based on instruction trace files[13], which yield the frequency of each instruction and histograms for the used operand fields. Typical ROM implementations store several memory words in one memory row according to Figure 6.5. Each access to one row results in toggle activity caused by all the stored words in this row. This fact requires redefining the instruction frequency as a basis for the above-mentioned optimization: An access to one specific memory row results in an access to several instructions in this model, thus, the access count of all the instructions residing in this row has to be incremented. In case of divided word line or split bit line memory implementations [176], this fact can readily be taken into account, by redefining the above-mentioned instruction frequency.

The decision, if it is worth storing a certain operand field (e.g. the immediate field of a MOVI instruction) using bit inversion, is based on the frequency of 1 bits and 0 bits in this field. If the frequency of 0 bits is higher than the frequency of 1 bits, this immediate field is inverted[14].

The decision, if an additional invert bit is to be used for a certain operand field, potentially increases the total instruction coding width. If the sum of the operand fields $W_{operand\,fields}$ of one instruction does not

[13] These instruction trace files need to be generated with the instruction set simulator.

[14] The value which is actually stored in the memory is the bit-inverted value of the original value. The original value has to be restored by bit-inversion in the ASIP decoder, based on the specific decoded instruction or based on the additional invert bit.

fully exploit the necessary instruction word length W_{instr}, this option is an alternative to the above-mentioned static bit inverted representation. Otherwise, this decision is left to the designer to trade-off toggle activity for instruction word width.

The opcode assignment is based on a maximum weight sorted set of all possible opcode instances assigned to the set of instruction types that are sorted according to the frequency in the instruction traces. A similar consideration applies to the swap operation of binary bits within the Huffman tree. The decision bits in the tree are swapped, if the 1 bit is in the branch with the lower frequency. The result of this optimization process is optimum for the given degrees of freedom, because of the optimality of the individual assignments, which individually maximize the number of 1 bits for each field.

Example: For the real-world ASIP of Section 7.1 the tool ICON has been applied in order to perform the above mentioned coding optimization. Table 6.1 depicts the results of this optimization. The unoptimized binary coding is an ad hoc-coding using prefix opcodes and logic 0s in the don't-care positions. A significant reduction of the toggle count can be achieved using ICON, which yields about 70% reduction of the internal bit line toggle activity. Depending on the percentage which is consumed by the toggle activity of the bit lines[15] overall power savings of 10% to 20% are achieved for the case study.

encoding technique	bit line toggles	savings (BL toggle count)
unopt. binary	2.93M	-
optimized encoding	0.829M	71,7%

Table 6.1: Results of Internal Memory Toggle Rate Optimization

In case of **off-chip instruction memory**, the optimization of the instruction coding is even more important, because of the impact on the toggle activity of the chip pad and external capacitances. The optimization problem in this case is different, because instead of maximizing the *static* number of 1 bits, the number of $0 \rightarrow 1$ and $1 \rightarrow 0$ transitions has to be minimized. The optimization problem can be expressed as

[15]Unfortunately, this percentage is unknown for the implemented memory of the case study in Section 7.1. In this case, we assume a percentage between 30% and 60% as mentioned above.

$$\arg_{IT} \min \sum_{j=0}^{j=W_{instr}-1} \sum_{i=1}^{i=N_{trace\_length}-1} IT_{i,j} \oplus IT_{i-1,j} \tag{6.6}$$

where $IT_{i,j}$ represents the j-th bit of the instruction i in the instruction trace of length N_{trace_length}. The exclusive-or relation can also be rewritten as

$$\arg_{IT} \min \sum_{j=0}^{j=W_{instr}-1} \sum_{i=1}^{i=N_{trace\_length}-1} IT_{i,j} + IT_{i-1,j} - 2IT_{i,j}IT_{i-1,j} \tag{6.7}$$

The instruction coding for Equation 6.7 obviously has to meet additional constraints like unambiguous opcode and operand field assignments, which have to be respected during the optimization process.

The degrees of freedom for this optimization are

- assignment of don't care bits
- use of non-inverted or inverted operand fields or redundant representation with invert bit
- assignment of opcode codings
- position of operand fields for each instruction type[16]

This optimization is difficult to handle, because of the huge problem complexity and the unavoidable overlap of different instruction fields. For this reason, the tool ICON uses a heuristic optimization technique, which is motivated in the following. Consider the case of a constant width opcode field, which occupies a certain bit range in the instruction. Without a loss in generality, we assume that this range occupies the bit positions 0 to $W_{opcode,enum} - 1$. Thus the task of optimizing the opcode assignment for this case has to minimize the expression

$$\sum_{j=0}^{j=W_{opcode,enum}-1} \sum_{i=2}^{i=N_{trace\_length}} IT_{i,j} + IT_{i-1,j} - 2IT_{i,j}IT_{i-1,j} \tag{6.8}$$

[16]The position of the opcode field itself has to start at a fixed location in order to enable the decode operation. This is equally true for split opcode fields, which are not explicitly covered in this discussion.

which is only a function of the actual opcode field assignment. The transition matrix for these opcode fields $\mathbf{T}_{opcode}(i,j)$ is defined as the number of transitions between instruction type i and instruction type j, where $i,j \in \{0, N_{instr\ types} - 1\}$. This matrix can easily be computed using the relevant information of the instruction trace. The heuristic of opcode assignment starts by finding a maximum value[17] in the off-diagonal elements of matrix $\mathbf{T}_{opcode}$, which yields two instruction types i and j with the highest transition frequency. Two codings of width $W_{opcode,enum}$ with a Hammming distance of one are assigned to these instruction types i and j. The maximum value in matrix $\mathbf{T}_{opcode}$ is marked *assigned* and the heuristic continues by finding the next unassigned maximum value in the off-diagonal fields and assigns a coding with minimum Hamming distance[18]. If several assignments are possible, the tool selects the coding that minimizes the incremental toggle count considering a parameterizable number of other already assigned instruction types.
This Greedy algorithm continues until all instructions are assigned. The degrees of freedom in this algorithm are chosen randomly in order to prune the complexity of the algorithm[19]. This approach allows to reiterate this algorithm several times in order to find better solutions.

Similar heuristics have been used to optimize the remaining assignment problems, which are more complicated, due to the fact that the different operand fields and the opcode field with prefix coding have overlapping bit ranges. Furthermore, the position of operand fields for each instruction type introduces an additional degree of freedom[20]. Heuristics for the opcodes and the operand assignments are used as a basis for a genetic optimization algorithm. The reproduction function of this algorithm uses code exchanges both within one code assignments and between several instances of code assignments. Table 6.2 shows the optimization results for a real-world case study using the above-mentioned

[17] Generally, there is no single maximum value in matrix $\mathbf{T}_{opcode}$.

[18] If one instruction type which is associated to this maximum is already assigned, the heuristic assigns a coding to the other instruction types with minimum Hamming distance.

[19] Exhaustive optimization for a relevant problem size is impossible due to the algorithmic complexity. Optimization approaches, which take into account several degrees of freedom (and use significantly more runtime), have not been able to yield significantly better results.

[20] Commercial processors often use fixed positions for operand fields which occur in several instruction types. This methodology saves multiplexers in the decoder. On the other hand, the area and power contribution of this part of the decoder in our case studies in Appendix F clearly shows that this additional complexity for the considered word widths is negligible.

toggle optimization. If we assume a load capacitance between 1pF and 10pF of the external instruction bus, this optimization saves between 1.8 and 9mW in total system power for a 100MHz system clock. Compared to the power consumption of the ASIP in Appendix F, which is in the order of 20mW, this saving is significant.

encoding technique	absolute toggles/1E3	savings
adhoc assignment	521	0%
optimized	241	53.7%

Table 6.2: Results of Instruction Bus Toggle Rate Optimization

Once the code assignments of all the instructions is finished, the task of generating the actual hardware decoder is straightforward. Currently, ICON uses VHDL as target description language. Listing 6.6 depicts extracts of the ICON-generated decoder description for a MIPS instruction set without coding optimizations. This decoder generation takes advantage of the capabilities of VHDL to use structured data types like records and enumeration types. This eases the development and the debugging of the code, because the instruction mnemonics instead of the instruction binary codes are shown in the waveform viewer during simulation .

6.3.2 Semi-Automatic Test Case Generation

In Section 5.4 the importance of hardware verification has been motivated. The proposed verification task can be facilitated using a test case generation (TCG) tool to automate the generation of test programs and test stimuli. This TCG tool has been conceived in order to provide stimuli for the cosimulation between a golden reference (which is typically a high level instruction set processor[21]) and a given hardware description. This simulation-based approach is comparable with the methodology typically used for commercial processor designs (cf. e.g. [68] [118] [149] [168] [174]) in order to cover typical fault models of the implementation.

[21] In [87] the equivalence between a VHDL implementation and a physical instance of the commercially available TMS320C25 processor has been verified.

```
entity score_predecoder is
port( clk, rstq: in std_logic;
insert_nop, insert_idle: in std_logic;
predecode_input: in std_logic_vector (22 downto 0);
predecode_output: out compl_operation_t);

end score_predecoder;

architecture predecode of score_predecoder is
signal predecoder_register: std_logic_vector (22 downto 0);
begin
process (predecoder_input)
begin
        -- opcodes
case predecode_input (22 downto 19) is
when "0000" =>
predecode_output.opcode <= addi_op;
when "0001" =>
predecode_output.opcode <= andi_op;
        ...
when others =>
  case predecode_input (22 downto 15) is
  when "11100000" =>
  predecode_output.opcode <= add_op;
  when "11110000" =>
  predecode_output.opcode <= sub_op;
                  ...
        -- operands
case predecode_input (22 downto 19) is
when "0000" | "0001" | "0010" | "0011" | "0100" | "0101" |
                    "0110" | "0111" | "1000" | "1001" | "1010" | "1011" |
                    "1100" | "1101" =>
predecode_output.reg0 <= conv_unsigned (0, 5);
predecode_output.reg1 <= unsigned ( predecode_input (18 downto 14));
predecode_output.reg2 <= unsigned ( predecode_input (13 downto 9));
predecode_output.imm <= signed ( predecode_input (8 downto 1));
predecode_output.addr <= conv_unsigned (0, 8);
predecode_output.raddr <= conv_signed (0,8);
when others =>
  case predecode_input (22 downto 15) is
  when "11100000" | "11100001" | "11100010" | "11100011" |
                      "11100100" | "11100101" | "11100110" | "11100111" |
                  ...
  end case;
end case;
end process;
end predecode;
```

Listing 6.6: Extract of a generated VHDL Decoder

If the abstraction level of operator-based RTL hardware design is used, the fault model for this verification task does not have to cover the operator implementations themselves[22]. The fault model for this verification

[22]This statement does not cover custom designed operator implementations but is rather valid, if operator implementations are taken from a synthetic operator library like the DesignWare library of Synopsys [233].

task rather focuses on the correct interconnections and the scheduling of the instanciated high level operators as well as on custom designed operators and on finite state machines.

Custom designed operators should be verified in a first verification step by e.g. exhaustive simulation or formal techniques. Exhaustive or nearly exhaustive simulation of these operators is very often possible, due to the limited amount of inputs for these operators. This verification task corresponds to the use of a divide-and-conquer verification methodology and helps to reduce the total amount of simulation vectors in order to obtain a certain coverage of the implementation.

The proposed TCG tool needs the user to evaluate the coverage of a given test program[23] suite and to define the basic structures of new test programs that enable a higher simulation coverage. Theoretically, this step can also be automated using an approach similar to [260] or [108]. Unfortunately, these approaches require a formal description of the hardware behavior and are restricted to simple architectures.

According to the results of this thesis that have already been published in [87], test case generation using pseudo-random test vectors typically achieves a remarkable percentage of state and execution coverage. This statement agrees with the results in [260]. The uncovered part of the design represent typically less than 5% to 10% percent and have to be covered using manual interaction.

The proposed TCG tool of this thesis is able to generate pseudo-random test programs and test I/O stimuli. Furthermore, the user can control this test vector generation with so-called *rules* to obtain a certain structure in the test program. By using these rules the user can describe e.g. program loops with a defined exit condition or similar conditional constructs. An example for these rules is given in Listing 6.7. These rules use a C-like description style in order to call a predefined ASIP-specific function (*gen_instr*) to generate an instruction. Instructions (e.g. *RPTK_INSN* for the C25 repeat instruction *RPTK*) and instruction groups (e.g. *ANY_REPEATABLE_INSN*) can be selected as argument for the function *gen_instr*. Furthermore, the arguments of these instructions can be constrained using e.g. the random function *RND*. The next

[23]Note that the term *test program* refers to the program which is actually used to *verify* the hardware description and not for hardware testing.

step is to speculatively simulate the generated instruction(s) and insert them - after successful simulation - in the test program (using the function *simulate_and_commit*). Other functions are able to generate labels (*gen_label*) and to generate random values that can be retrieved later on (*clear_arr*, *ST_RND* and *RE_VAL*).

```
/* rule to use a C25 repeat immediate instr. */
RPTK_RULE:
gen_instr(RPTK_INSN,RND(3,20));          // generate RPTI instr.
gen_instr(ANY_REPEATABLE_INSN);          // generate repeated instr.
simulate_and_commit(30);                 // speculative execution of max.
                                         // 30 cycles and commitement in
                                         // case of successful simulation

/* rule to generate a loop with constant iteration count (ICORE) */
LOOP_RULE:
gen_instr(LPINI_INSN,RND(2,10),PC+6);    // random repeat count
for(int i=0; i<=5; i++)                  // generate 5 random arith.
  gen_instr(ANY_ARITH_INSN);             // instructions with rand. operands
simulate_and_commit(60);                 // speculative execution of max.
                                         // 60 cycles and commitement in
                                         // case of successful simulation

/* rule to check, if BEQ works (ICORE) */
BEQ_RULE:
clear_arr();
gen_instr(MOVI_INSN,ST_RND(-128,127),"R"ST_RND(0,7));
                                         // generate MOVI instr. and
                                         // remember the operand values
gen_instr(CMPI_INSN,RE_VAL(0),"R"RE_VAL(1));
                                         // generate CMPI instr. using the
                                         // remembered operand values
gen_instr(BEQ_INSN,"BEQ_OK_LB");         // generate BEQ instruction
gen_instr(HALT_INSN);                    // halt on error
gen_label("BEQ_OK_LB'');                 // generate unique label
                                         // BEQ_LB_#xyz#
simulate_and_commit(10);                 // speculative execution of max.
                                         // 10 cycles and commitement in
                                         // case of successful simulation
```

Listing 6.7: Example Rules for Test Generator

According to these rules, the TCG tool can generate specialized instruction sequences that are necessary to produce valid DSP assembler code[24] even for unorthogonal architectures. Furthermore, the test case generator has to meet *constraints* of the target architecture in order to generate valid code. Examples for these constraints are register file and stack sizes, valid memory spaces or restricted parameter ranges for cer-

[24]For the TMS320C25 specialized instruction sequences are needed to verify the RPT-instruction and some arithmetic instructions like SUBC.

tain arithmetic instructions. The degrees of freedom for the instruction generation that are not restricted by the generation rules and the constraints are determined using a pseudo-random generator. So-called *biases* using non-uniform probability functions increase the portion of the generated code that stresses error-prone corner cases. Examples for these corner cases are compare results equal/unequal to zero and arithmetic overflows.

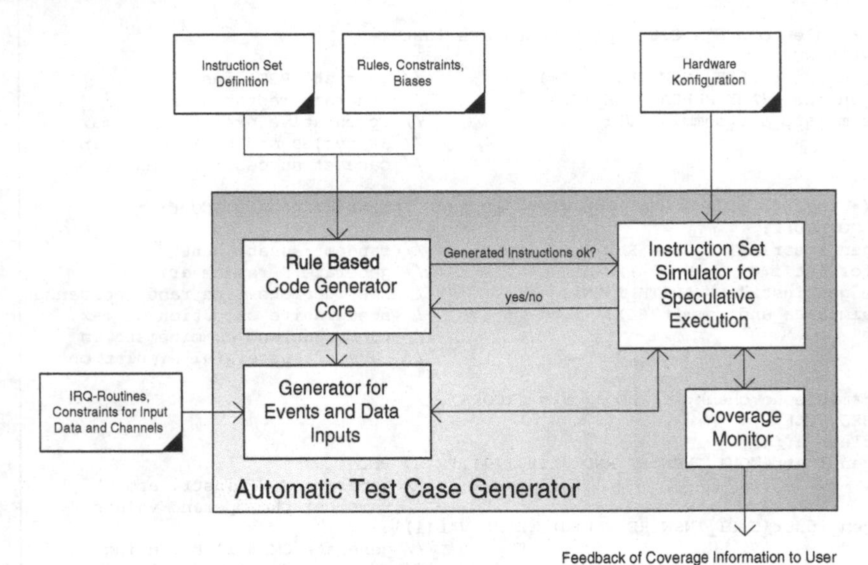

Figure 6.7: Structure of the Automatic Test Case Generator

Figure 6.7 depicts the structure of this test case generator. Due to the fact that the rules can leave many degrees of freedom to the pseudo-random code generator, it is not guaranteed that the initially generated code meets the constraints of the target architecture. In order to solve this problem, the generated instructions and instructions sequences are speculatively simulated with an instruction set simulator to check the validity of these instructions (e.g. to avoid memory access violations, incorrect operand values etc.). If a certain instruction/instruction sequence is valid, it is appended to the final test program and the next instruction/instruction sequence is generated. This concept enables the test program generator to assess the state coverage of the generated test program and immediately provide feedback to the user, who can modify the rules and biases in order to enhance the coverage. The test generator is also able to generate stimuli for the input channels of the processor as well as external events like IRQs, suspend and resume signals. After a suite of test programs and stimuli have been generated, the HDL description of the processor has to be cosimulated with the golden reference. In order to debug the HDL description either a cosimulation interface or generated trace files can be used, which trace the results of program tasks or the complete program flow with all relevant states.

6.4 Concluding Remarks

This chapter has briefly introduced the concept of the LISA language and the important software and hardware design development tools that can be generated using this description. ASIP-specific extensions to this tool suite have been introduced covering the optimization of the instruction encoding in order to save energy and the generation of the hardware decoder. With this methodology significant savings in energy and design time have been achieved for the real-world ASIP case study in Section 7.1.

Furthermore, a test case generator (TCG) to support and facilitate the generation of ASIP test programs and test stimuli has been presented. This tool partially automates the tedious process of generating test programs and helps to reduce the design time while enabling a higher simulation coverage. This TCG provides a transparent C-like script language as user interface in order to generate meaningful programs to stimulate error-prone parts of the design.

In order to transform the LISA ASIP design environment into a design platform for ASIPs a comprehensive library of processor templates and example processor designs for various application domains has to be implemented. These library elements should include LISA descriptions and optimized HLL compilers as well as fully verified hardware descriptions. The availability of these templates and examples will be a key factor to speed up iterative instruction set optimization for many applications.

Chapter 7

Case Studies

In this chapter the results of two case studies are presented to prove the feasibility of the proposed design flow. The first case study is about an acquisition and tracking control processor for terrestrial digital video broadcasting (DVB-T) and demonstrates the impact of ASIP optimizations on energy-efficiency. The second case study covers an ASIP for linear algebra kernels with a special focus on eigenvalue decomposition of hermitian matrices. This case study also compares the design and implementation efficiency of an optimized ASIP implementation with a general purpose processor core.

7.1 Case Study I: DVB-T Acquisition and Tracking

This case study is about the design of an ASIP that controls the acquisition and tracking process in a DVB-T receiver. Figure 7.1 depicts the simplified structure of this receiver. The DVB-T standard [75] uses coded orthogonal frequency division multiplex (COFDM) as a transmission technique. The term *coded* in this context means, that the transmitted data are protected against transmission errors by using convolutional and block codes. *Orthogonal frequency division multiplex* on the other hand means, that the transmission channel in the frequency domain is subdivided into equidistant subchannels. This principle can be viewed as a modulation of many equidistant carriers with a corresponding data sequence. The realization of this transmission technique uses the inverse discrete fourier transformation (IDFT – implemented by the inverse fast fourier transformation) for modulation and the discrete fourier transformation (DFT) for demodulation. In order to avoid inter-symbol interference (ISI) of two consecutive OFDM symbols, a guard interval is inserted that has to be longer than the duration of the channel impulse response. Frame synchronization in the receiver with respect to the position of the guard interval is needed to reduce the effective inter-symbol interference. Frequency synchronization before the DFT in the receiver mitigates the effect of a sampling frequency offset between the transmitter and receiver. The signal after the DFT in the receiver corresponds to the original signal before the IDFT multiplied with the DFT of the channel, provided that the frame synchronization and the frequency synchronization in the receiver are perfect. For DVB-T, phase correction after the DFT is performed with pilot carriers, because each subchannel uses a QAM modulated carrier which is sensitive to phase errors.

The underlying application of this section computes the acquisition of the FFT window position (timing synchronization), the sampling clock synchronization (interpolation/decimation control) and the carrier frequency offset estimation (frequency synchronization) [89]. Furthermore, after the acquisition is finished, the common phase error, the frequency error and the sampling error are continuously tracked. This lock condition is permanently monitored and automatic reacquisition is performed if needed.

The implementation for this DVB-T acquisition and tracking (DVB-T A&T) application has been named *ICORE*[1]. The final ICORE implementation including the hard- and software is integrated as one design module into a commercial single chip. This system-on-a-chip solution for DVB-T supports enhanced algorithms and features compared to the previous receiver generation [123].

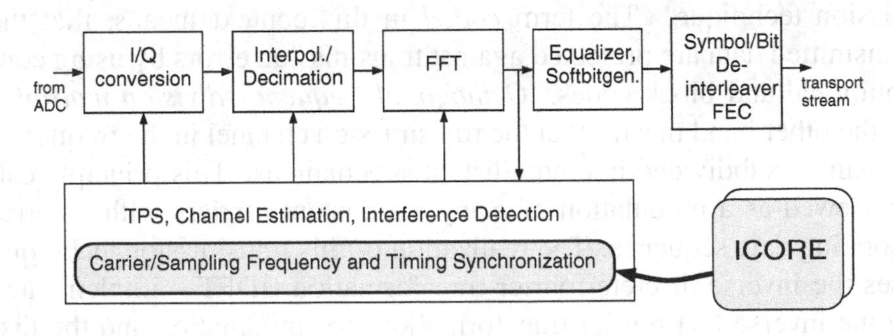

Figure 7.1: Digital Part of the DVB-T Receiver

[1]ICORE is the abbreviation for ISS-Core.

7.1.1 Application Profiling and ASIP Class Selection

The results of application profiling are briefly summarized in this subsection. For a more detailed description of this design task refer to Subsection 5.2.2, where the DVB-T A&T application is used as a vehicle to illustrate the profiling methodology.

Application profiling reveals the following results:

- cycle count violations of the profiling software implementation
- high data locality
- many data dependent branches (which are difficult to predict)
- many bit-oriented operations and arithmetic saturation operations
- insignificant part of regular DSP operations like e.g. FIR filters
- long idle intervals
- the time critical tasks frequently need arctan-computations

These application profiling results are used in order to determine a suitable ASIP class as a starting-point for further optimization. The high data locality of the application suggests a typical load/store architecture with a general purpose register file. A short processor pipeline is advantageous, in order to decrease the penalty of stall cycles due to the high number of branches. Generally, simplicity of the hardware is preferred over complexity to reduce the design and verification effort. This approach also enhances the maintainability and reusability of this design, which are two of the primary design goals together with a high energy-efficiency.

In order to find the optimum pipeline organization, several implementations with a different pipeline length and organization have been implemented for the DVB-T A&T application[2]. A detailed overview of the pipeline organization for the different alternatives is given in Appendix D. Table 7.1 depicts the results of this design space exploration for the DVB-T A&T application benchmark. The silicon area displayed

[2]These implementations have been separately optimized for runtime and in order to obtain a minimum in energy consumption.

in Table 7.1 increases and the critical path decreases with an increasing number of pipeline stages, as expected. For the overall runtime, the number of cycles has been multiplied with the critical path of the implementation. Due to data and control hazards reducing the average resource utilization in the ASIP pipeline, the change in absolute runtime is less than proportional to the change in the critical path for an increasing number of pipeline stages. Interestingly, a minimum in energy consumption is observed for the three-stage pipeline implementation.

Further investigations reveal that the two-stage implementation needs more energy due to logic glitches caused by a significant signal slack. The three-stage implementation reduces this effect by retiming of unbalanced signal arrival times in the additional pipeline stage. Moreover, slower and smaller arithmetic operator implementation are instanciated in the three-stage implementation due to the more relaxed timing constraints. The four-stage implementation needs more energy than the three-stage implementation in the clock circuitry and the flip-flops for the pipeline registers as well as in the hazard detection and resolution logic. For the three- and four-stage implementations, which use a predict-untaken branch scheme, taken branches result in a branch penalty of 2 and 3 cycles resp. This branch penalty results in a higher energy consumption of the four-stage implementation due to redundant additional fetch and decode operations.

For the DVB-T receiver system, the two stage implementation violates the given clock cycle constraint of the system environment. Furthermore, the four stage architecture requires higher effort for verification and design due to the hazard detection and resolution logic. The flat minimum in energy consumption is another reason, that the three-stage implementation has been the final architecture of choice for the DVB-T A&T application.

# of pipeline stages	2	3	4
norm. area	100%	103%	120%
norm. crit. path	100%	83%	72%
norm. benchmark runtime	100%	92%	86%
norm. energy	100%	85%	106%

Table 7.1: Results for Different Pipeline Structures

It has to be pointed out, that the comparison in Table 7.1 uses the best implementations w.r.t. energy consumption including all possible power optimizations of the next subsections for each case. For the sake of conciseness, the discussion of these optimizations in the following is restricted to the three-stage implementation.

7.1.2 Iterative Instruction Set Optimization

The purpose of typical ASIP instruction set optimizations is to enhance the computational performance. If the performance goals are reached, additional optimizations can be applied in order to increase the energy-efficiency of an application. The following two examples illustrate the effect of these instruction set optimizations. The first example is an optimized instruction performing the saturation of an integer value to the number range of a 2's complement number with programmable bit width. The second example is a CORDIC computation in vectoring mode, which uses several highly optimized instructions.

7.1.2.1 Example 1: Saturation

In the instruction traces of the profiling implementation more than 12% of the total executed instructions are used for saturation to a power of 2. This kind of saturation is defined by the following simple relation

$$\mathrm{sat}(n,m) = \begin{cases} 2^n - 1 & \text{if} \quad m > 2^n - 1 \\ -2^n & \text{if} \quad m < -2^n \\ m & \text{else} \end{cases}$$

Many commercially available DSPs offer a so-called saturation mode. This mode is usually restricted to saturate accumulator values with a long bit width to the shorter data path width of the DSP. However, the above-mentioned saturation task, saturates to any valid bit width from 1 bit to the full data path width. This property is required by the given DVB-T A&T algorithm in order to guarantee, that a certain range for critical output and intermediate values is not exceeded and malfunctions due to wrap-around are avoided.

Two implementation alternatives are considered for this task (cf. Figure 7.2):

- a pure software implementation[3] that uses the basic profiling instruction set of Table 5.4
- a specialized instruction SAT(bw,Rn), which saturates the register Rn with a dedicated functional unit in hardware to the minimum and maximum values of a 2's complement number with bit width bw

Optimized SW Implementation

```
SATURATE(15,R3);
```

SW Implementation with Profiling Instruction Set

```
MOV(R3,R2);
MOVI(#15,R0);
BSR(SATURATION);
MOV(R2,R3);
```

Subroutine Call

```
SATURATION: (expects exponent in R0 and value for sat. in R2)
  MOV(POS_VAL_TABLE+R0,AR0);
  RDM(AR0,R1);
  CMP(R1,R2);
  BLE(SAT_L1);
  MOV(R1,R2);
  RTS;
SAT_L1:
  MOV(NEG_VAL_M+R0,AR0);
  RDM(AR0,R1);
  CMP(R1,R2);
  BGE(SAT_L2);
  MOV(R1,R2);
SAT_L2:
  RTS;
```

Figure 7.2: Two Implementations for Programmable Saturation

The specialized SAT instruction is executed in a single instruction cycle without increasing the critical path of the implementation, whereas the conventional solution needs an average of 14.9 cycles in the benchmark

The result of the power and energy evaluation is given in table 7.2. The average power of the profiling implementation is about the same as the power of the optimized one. However, the results clearly show that the optimized implementation is far superior in energy-efficiency and

[3]For this implementation, a subroutine has been used in order to save program memory. Furthermore, in order to avoid the computation of the limits, a look-up-table is needed for this implementation.

Implementation	Unoptimized Profiling ISA	ISA with specialized SAT instruction
avg. cycles per saturation	14.9	1
avg. power [mW] (only sat.)	9.8	10.9
avg. Energy [nJ] per saturation	1.83	0.136
Relative avg. energy per saturation	100%	7.4%

Table 7.2: Results of the Saturation Benchmark

performance, because of the significant cycle count reduction. If additional spill code had been necessary to avoid overwriting of register values by the subroutine, the energy consumption of the unoptimized version would even have been increased due to more instructions and several RAM accesses.

Other implementations of the saturation task are also possible e.g. inlining of the code using constants for the limits, calculation of the max/min-values on the fly etc. However, all of these implementations require more program memory and more instructions than the optimized implementation resulting in a significantly lower energy-efficiency.

7.1.2.2 Example 2: CORDIC

The DVB-T A&T application require the CORDIC algorithm in vectoring mode to calculate the angle between a 2-dimensional vector (x, y) and the x-axis ($atan(x/y)$). This CORDIC task requires a significant amount of runtime in the time critical tasks of the profiling implementation, and, therefore, is a candidate for thorough optimization.

The time and power consuming CORDIC loop body has been implemented with differently specialized instructions according to Figure 7.3:

- hand-programmed implementation with the profiling instruction set (denoted *Implementation 1* in Figure 7.3 and Table 7.3) without zero-overhead-loop instruction
- specialized instructions (including zero-overhead loop support) using special purpose hardware units like shift-and-round[4], conditional addition/subtraction etc. (*Implementation 2*)
- even more specialized instructions and a second addition/subtraction-unit in the core (*Implementation 3*)

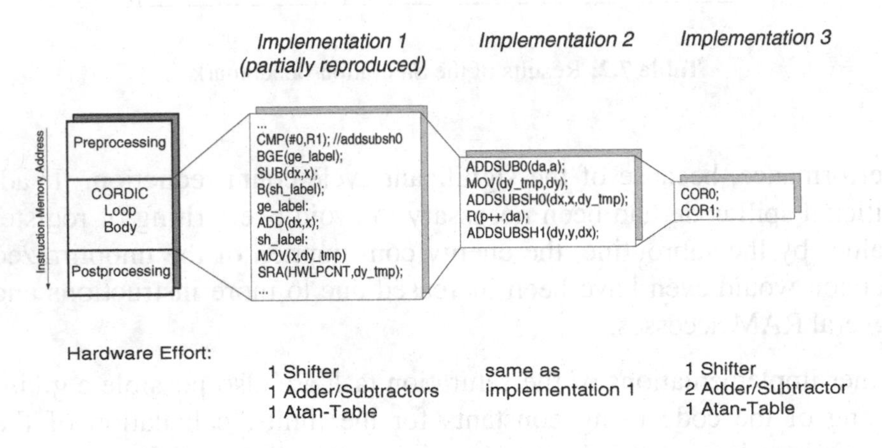

Figure 7.3: Implementation Alternatives for CORDIC Loop Body

In table 7.3 results of the CORDIC task evaluation are depicted. The average energy consumption of this task is normalized to the profiling ISA implementation.

Implementation	Impl. 1	Impl. 2	Impl. 3
avg. cycles per CORDIC call	663.3	154.8	82.4
relative avg. power	100%	115%	127%
relative avg. energy per CORDIC call	100%	18.8%	15.8%

Table 7.3: Results of the Different CORDIC Implementations

[4]The CORDIC task for implementation uses rounding of the LSB after shifting. An alternative to this algorithm is to use at least one additional fractional bit.

Table 7.3 clearly shows the effect of instruction set specialization on the average power consumption: Due to the parallel execution of operations in the more specialized implementations, the average power increases. However, the decrease in runtime of the CORDIC task overcompensates this increase and results in significant energy savings. Table 7.3 shows, that the best optimized version of the CORDIC consumes about 6.7 times less energy than Implementation 1 (profiling instruction set).

7.1.3 Overall Energy Optimization Results

Instruction set optimizations typically decrease the runtime of the processor for a given task. In order to continue increasing the energy-efficiency of an implementation, additional instruction set optimizations can be applied, even if the runtime constraints of a given application are already met. Apart from ISA optimizations, additional architectural optimizations can be used that have been described in Section 3.3.

This subsection summarizes the energy optimization results of the ICORE three-stage implementation. The numbers in Figure 7.4 are related to the effect of incremental optimizations beginning with an implementation using the profiling instruction set of Section 5.2.2. These optimizations are sorted starting with the least effective optimization (logic restructuring) and ending with the most effective ones (clock gating and instruction set optimization). It is important, that clock gating has to be introduced before instruction set optimization, otherwise, the benefit of a longer processor sleep period due to faster processing is significantly reduced.

Reorganization of logic gates and operators[5] is the least efficient optimization yielding about 10% in energy reduction, but without significant increase in design time. **Blocking gates**[6] reduce the energy by roughly another 10% while increasing the area by less than 1%. The reduction of the internal instruction ROM toggle activity yields about 20% in energy reduction using automatic **optimized encoding**[6] without affecting area or design time. This saving depends on the size and organization of the instruction memory. Application-specific **instruction**

[5]Refer to Subsection 3.3.2 for a description of this energy saving technique.

[6]This technique has been developed and published in [129] [90] and is based on the tool which is described in Subsection 6.3.1

set optimization[7] cuts energy consumption by another 50%, while increasing the design effort significantly due to manual optimization. The applicability of this optimization strongly depends on the computational tasks of the application. The benefit of **clock gating** in combination with the sleep mode of the core yields a factor of about four in energy reduction, because the processor for the DVB-T A&T application has long idle intervals. This value strongly depends on the workload of the ASIP. It might be argued that the processing power of ICORE is significantly over-dimensioned for the given application. This is not the case: The runtime constraints of the application rather represent tight bounds for the ICORE tasks that have to be met by this implementation.

The overall power reduction for the three-stage ICORE implementation with all the above-mentioned optimizations is about 92%. It has to be pointed out that all these optimizations do not compromise the flexibility and maintainability of this building block for late design changes that require the implementation of additional software programmable tasks.

In the following discussion, the implementation of an ASIP accelerator for the computationally intensive CORDIC task is explored. This implementation is similar to the example of Subsection 5.3.7, but in this case stripped down to a CORDIC for vectoring mode. This coprocessor has been implemented and connected to the ASIP core. Table 7.4 shows the area and energy consumption for the different implementations (ASIP with/without coprocessor) for the CORDIC task and, additionally, for the overall DVB-T A&T benchmark tasks (which include several CORDIC evaluations). The overall savings for the complete tracking tasks are about 38%.

Nevertheless, the implementation of an accelerator breaks the design paradigm of an instruction set oriented ASIP and introduces more heterogeneity into the implementation. Maintainability and reusability of this building block become more complicated, which outweighs the additional gain in energy-efficiency for the DVB-T system. Consequently, the final ICORE has been implemented using the above-described software CORDIC implementation.

[7] Refer to the previous subsection for an example.

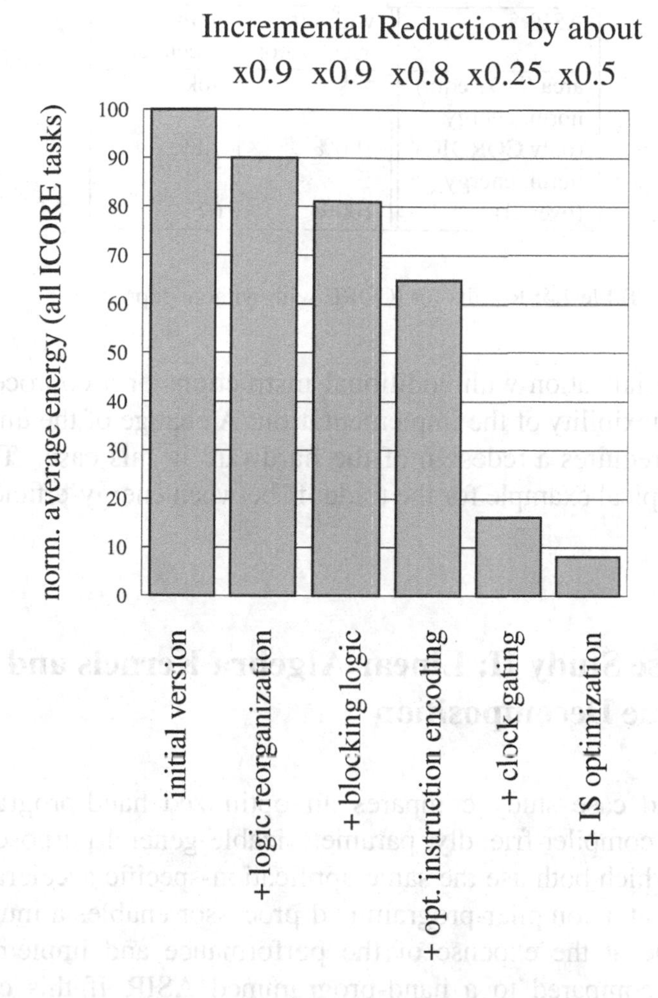

Figure 7.4: Incremental Power Optimization of ICORE

Generally, it can be observed that the energy-efficiency of an implementation increases, if specialization for a given application is introduced. The above-mentioned instruction set optimization and the implementation of accelerators represent an *incremental* modification resulting in increased energy-efficiency. At the same time the overall flexibility and reusability of the processor for unexpected tasks with low to medium required computational performance is preserved. For the performance critical computational tasks like the CORDIC computation mentioned

ASIP	without . accelerator	with accelerator
area (ND2 equ.)	52k	56k
norm. energy (only CORDIC)	100%	7.8%
norm. energy (overall)	100%	62%

Table 7.4: Results for ICORE with/without coprocessor

above, specialization with additional instructions or a coprocessor reduces the flexibility of the implementation: A change of the underlying algorithm requires a redesign of the hardware in this case. The latter case is a typical example for the tradeoff between energy-efficiency and flexibility.

7.2 Case Study II: Linear Algebra Kernels and Eigenvalue Decomposition

The second case study compares an optimized hand-programmable ASIP to a compiler-friendly, parameterizable general purpose processor core, which both use the same application-specific accelerator. It is obvious, that a compiler-programmed processor enables a much faster design time at the expense of the performance and implementation efficiency compared to a hand-programmed ASIP. If this compiler-programmable processor can be parameterized in order to match the performance requirements of an application, however, this concept is useful for many applications.

The target application are linear algebra kernels for communication applications. These kernels include typical complex matrix and vector/matrix operations (cf. Appendix B.4). The specific benchmark for this case study is the eigenvalue/eigenvector decomposition (EVD) of a hermitian matrix[8] using a Givens-like decomposition algorithm (cf. Appendix B.5). Complex linear algebra and matrix decomposi-

[8] However, the instruction set of the presented ASIP is not restricted to the EVD but also supports singular value decomposition of rectangular matrices and the vector-matrix operations mentioned above.

tion techniques are needed for various communication applications that use subspace decomposition e.g. direction of arrival (DOA) estimations [219] [151], beamforming [6], adaptive filter processing [105] and vector quantization [279].

The two different design approaches that are compared in this chapter are

- an optimized ASIP tailored to the given application by using specialized instructions and a specialized data path together with a dedicated accelerator ("constructive ASIP design methodology", which has been described in Section 5.3.6)
- a processor core with a fixed instruction set, but a parameterizable number of functional units like multipliers, adders and memory units together with the same dedicated coprocessor as above ("pure library-based ASIP design methodology")

Both methodologies take advantage of the concept of a tightly coupled coprocessor. For this case study, the dedicated CORDIC coprocessor that has already been described in the example of Subsection 5.3.7 is used. This coprocessor is able to significantly reduce the computational load of both processors by mapping a regular computational part of the overall algorithm to dedicated hardware.

The two implementations are described in detail in the following two subsections. Afterwards in Subsection 7.2.3, the evaluation of these implementations is presented using the eigenvector and eigenvalue decomposition of a 10x10 hermitian matrix as benchmark application.

7.2.1 Implementation I: Optimized ASIP with Accelerator

The algorithm for Givens-like eigenvalue and eigenvector decomposition is described in Appendix B.5. In addition to the trigonometric functions like sine, cosine, phase and magnitude of a vector, the EVD requires matrix-matrix-multiplications. These multiplications with the Givens-matrix $\mathbf{G}_n$ are used to iteratively update both the matrix being diagonalized, $\mathbf{A}_n$, and the matrix containing the approximate right (left) eigenvectors, $\mathbf{EV}_{r(l),n}$. Figure 7.5 depicts this matrix-matrix-multiplication for the right eigenvector matrix update using a 4x4 matrix with the Pivot element $(2, 4)$ as example

$$
\mathbf{EV}_n \cdot \mathbf{G}_n = \begin{pmatrix} e_{11} & e_{12} & e_{13} & e_{14} \\ e_{21} & e_{22} & e_{23} & e_{24} \\ e_{31} & e_{32} & e_{33} & e_{34} \\ e_{41} & e_{42} & e_{43} & e_{44} \end{pmatrix} \cdot \begin{pmatrix} 1 & 0 & 0 & 0 \\ 0 & q_{11} & 0 & q_{12} \\ 0 & 0 & 1 & 0 \\ 0 & q_{21} & 0 & q_{22} \end{pmatrix}
$$

$$
\mathbf{EV}_{n+1} = \begin{pmatrix} e_{11} & q_{11}\,e_{12}+q_{21}\,e_{14} & e_{13} & q_{12}\,e_{12}+q_{22}\,e_{14} \\ e_{21} & q_{11}\,e_{22}+q_{21}\,e_{24} & e_{13} & q_{12}\,e_{22}+q_{22}\,e_{24} \\ e_{31} & q_{11}\,e_{32}+q_{21}\,e_{34} & e_{13} & q_{12}\,e_{32}+q_{22}\,e_{34} \\ e_{41} & q_{11}\,e_{42}+q_{21}\,e_{44} & e_{13} & q_{12}\,e_{42}+q_{22}\,e_{44} \end{pmatrix}
$$

left column | right column

Figure 7.5: Eigenvector Matrix Update

For a real-world implementation, one multiplication in the Equations B.12 and B.13 is immediately performed, after the Givens-matrix $\mathbf{G}_n$ is available, rather than storing the matrices $\mathbf{G}_n$ and postponing this calculation. Due to the fact that the Givens-matrices $\mathbf{G}_n$ represent the identity matrix with the embedded 2x2 pivot submatrix $\mathbf{Q}_n$, the matrix-matrix-multiplication reduces to a multiplication with the matrix $\mathbf{Q}_n$.

In contrast, the update of the matrix $\mathbf{A}_n$ can exploit additional arithmetic simplifications due to the symmetry of the hermitian matrix $\mathbf{A}_n$, which results in a significantly reduced number of arithmetic operations.

A critical issue of high performance signal processing is memory organization, because memories often represent a bandwidth bottleneck due to a small number of read and write ports (typ. 1 or 2 write ports are available). The degree of parallelism in the EVD algorithm is in the order of the matrix dimension, which theoretically enables a full parallel solution with distributed parallel memory blocks or even registers. Unfortunately, the hardware costs of this solution in terms of silicon area are proportional to the matrix dimension and do not justify the benefit in computational performance of this massive parallel approach for the given application constraints. In the current case study, the structure of the matrix updates suggests a dual-port RAM as main data memory for matrix computations, because 2 samples are processed in each computation step. This design decision represents a trade-off between a full parallel and a scalar implementation.

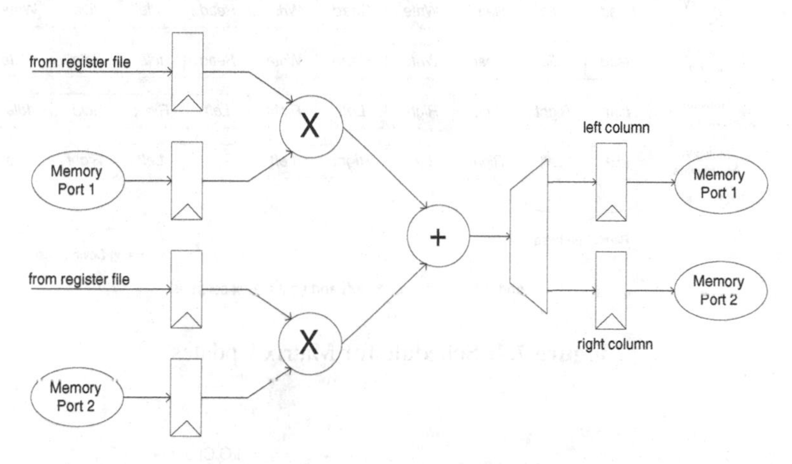

Figure 7.6: Computational Core for Matrix Updates

A simple computational core that enables the necessary functionality for the matrix update functionality is depicted in Figure 7.6. This core needs registers for pipelining and to reduce memory accesses by storing the frequently needed values in $\mathbf{G}_n$ in a register file. The eigenvector matrix $\mathbf{EV}_n$ is stored in the dual-port RAM. A schedule for the eigenvector matrix update is depicted in Figure 7.7, which demonstrates the efficient use of the computational resources and the dual-port memory. An extended version of the computational structure in Figure 7.6 is used as part of the vector functional units in the final implementation of this case study.

This final architecture of the ASIP has been named *ICORE-II* and supports both scalar and vector instructions that match the properties of the application. Figure 7.8 depicts a simplified overview of the important parts of ICORE-II: The main difference to the scalar ICORE are the vector functional units and the parallelized data memory. The vector functional units are tightly coupled to the scalar part of the core by sharing the general purpose register file in order to save area and for efficient communication. Furthermore, the decoder generates the parallelized control information by using an additional vector decoder, which is in turn controlled by a microcode sequencer. This sequencer supports multi-cycle instructions and controls the processing for vector lengths that exceed the available parallelism in the hardware.

Memory Port 1	*Read*	*Idle*	*Read*	*Write*	*Read*	*Write*	*Read*	*Idle*	*Idle*	*Write*
Memory Port 2	*Read*	*Idle*	*Read*	*Write*	*Read*	*Write*	*Read*	*Idle*	*Idle*	*Write*
Register Read	*Left*	*Right*	*Left*	*Right*	*Left*	*Right*	*Left*	*Right*	*Idle*	*Idle*
Multiply-Accumulate	*Idle*	*Left*	*Right*	*Left*	*Right*	*Left*	*Right*	*Left*	*Right*	*Idle*

Ramp-Up Phase

Ramp-Down Phase

"Left" and "Right" refer to the left and right matrix columns

Figure 7.7: Schedule for Matrix Updates

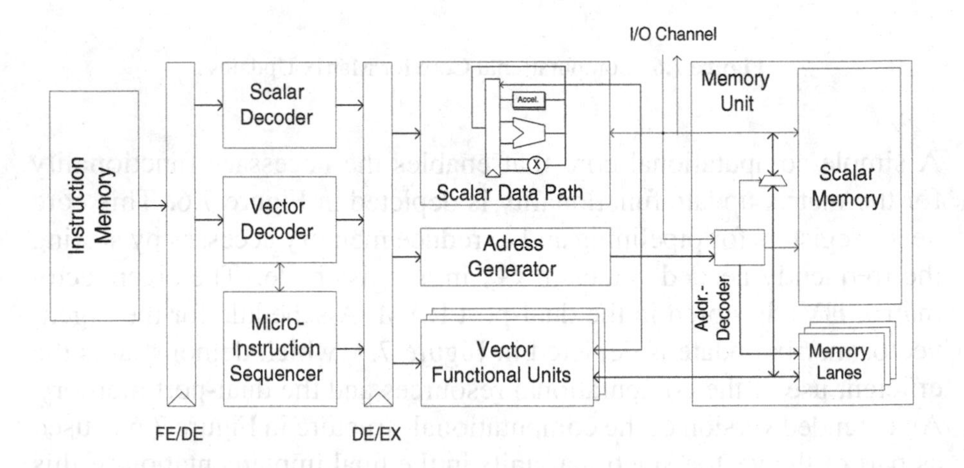

Figure 7.8: Simplified Overview of ICORE-II

ICORE-II supports general vector instructions with a programmable vector length for scaling, dot-products and matrix-matrix additions and multiplications. Additional application-specific instructions and addressing modes have been implemented both for the scalar and the vector part of the implementation addressing the EVD application:

- instructions to support the CORDIC coprocessor
- addressing modes to access row- and column-indexed matrix elements residing in the memory address $base\_address + row\_register * dim\_register + col\_register$
- instructions to support the update operation for the matrix **A** and the eigenvector matrix **EV**

7.2.2 Implementation II: Compiler-Programmed Parameterizable Core with Accelerator

The considered parameterizable processor core of this section has been named *ALICE*[9] [271]. The architecture of this core, which is depicted in Figure 7.9, uses 5 pipeline stages (FEtch, DEcode1, DEcode2, EXecute and WriteBack).

In the stage DE2 the general purpose register is read and the register output is routed to the input registers of the functional units in the EX stage. Due to the fact that the number of functional units is parameterizable, the necessary bandwidth for the control information has to be provided by an equally scalable instruction fetch stage. For the ALICE architecture, the number of fetch lanes has to be a power of two in order to enable a simple addressing logic. There is one program memory lane and a lane decoder associated to each fetch lane. Instructions can be fetched in parallel using the concept of compressed VLIW encoding, which has been explained in Section 4.3.7. Thus, "no-operation" instructions do not have to be explicitly coded and the associated program memory locations can be saved for useful instructions. Figure 7.10 shows an example program, which takes advantage of the parallelism provided by ALICE using 4 fetch lanes. The parameters of ALICE that can be adjusted to the needs of an application are

- the word width of the data path

[9]Architecture for LISA Compiler Environment

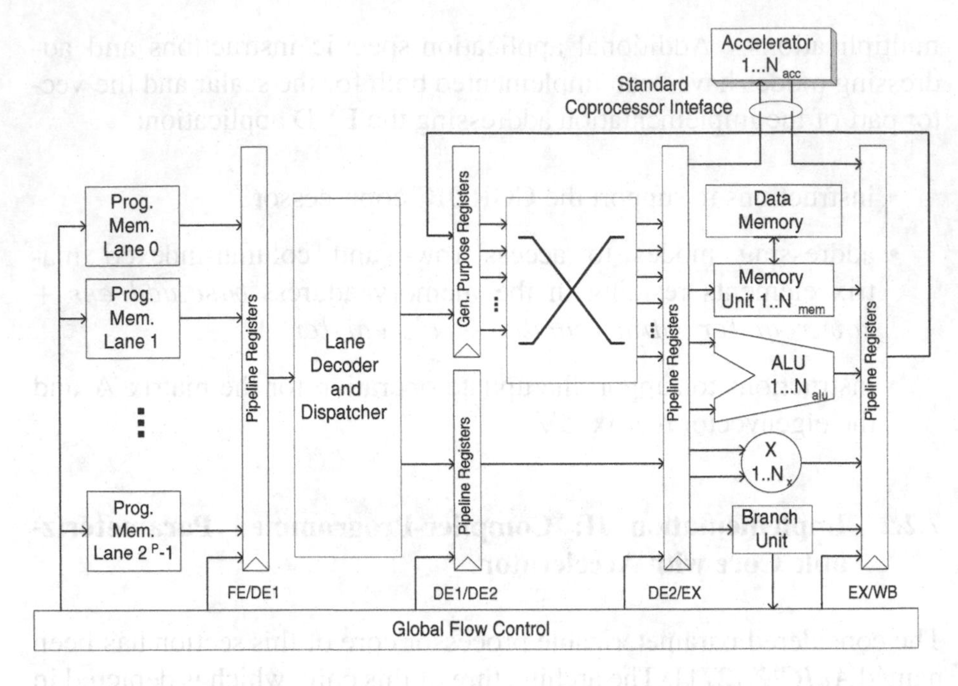

Figure 7.9: Overview of ALICE Architecture

- the number of parallel instruction memory lanes
- the number of general purpose registers and the number of read/write ports
- the forwarding configuration: enable/disable forwarding from WB to EX and from WB to DE2
- the number of functional units including arbitrary accelerators and memory units
- the branch behavior configuration: enable/disable branch delay slots

The user has to take care to select a reasonable configuration that considers the mutual dependencies between some of the parameters mentioned above: e.g. it does not make sense to instanciate several ALUs without providing sufficient instruction memory bandwidth by increasing the number of instruction memory lanes appropriately.

Mem.-Line	Lane 1	Lane 2	Lane 3	Lane 4
0	ADD	SUB	LW	MUL
1	SLL	SLTI	MUL	OR
2	SW	BEQ	MUL	MUL
3	J	ADDI	SW	...

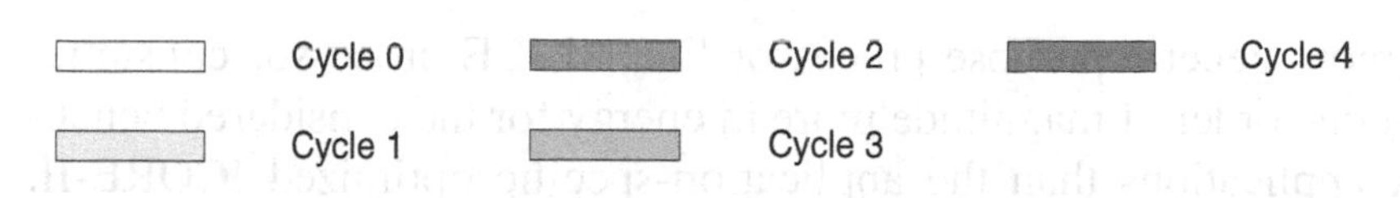

Figure 7.10: Example ALICE Program in Memory

The question arises, if there is a difference between ALICE and the concept of a general ASIP. Indeed, there is one important distinction: For the current case study the instructions of ALICE are fixed, which means that the user does not have to care about micro-architectural details and optimizations[10]. As a replacement of ISA optimization, ALICE supports a wide variety of configuration parameters, in order to scale the available architectural parallelism. Furthermore, the use of accelerators with well-defined hardware/software interfaces[11] in the core enable to take advantage of the high efficiency of dedicated hardware for regular arithmetic computations. Finally, ALICE represents an orthogonal ISA, which can be efficiently targeted by HLL compilers (cf. [271]) to speed up the design and verification process.

7.2.3 Evaluation Results

The two processor cores ICORE-II and ALICE have been implemented in VHDL and synthesized using a typical 0.18μ technology. Power estimation has been performed with Synopsys' DesignPower using the toggle information of gate level simulation. The results for these two

[10] It is obviously still possible to add optimized applications specific instructions to ALICE later on in the design flow. However, the primary goal of the current case study is to avoid this optimization for ALICE in order to evaluate the reduction in design time.

[11] The software interface is realized with fixed instructions similar to the concept of the ARM [16] or SPARC [228] instruction set with reusable accelerator move and control instructions.

implementations with the EVD as benchmark application are depicted in Table 7.5. In this configuration, ALICE uses 4 parallel fetch lanes, 2 ALUs, 2 multipliers and the same CORDIC coprocessor than ICORE-II as a special purpose unit.

According to Table 7.5 the generic scalability of ALICE significantly increases the implementation area compared to the application-specific ICORE-II implementation.

The more general purpose processor like ALICE processor consumes about one order of magnitude more in energy for the considered benchmark applications than the application-specific optimized ICORE-II. This is due to the larger instruction memories for parallel instruction fetch, the significantly larger register file, as well as the additional logic for instruction expansion and instruction dispatch. Furthermore, ALICE is deeply pipelined requiring many power intensive pipeline registers and forwarding paths. This difference in energy consumption is a classical example for the tradeoff between energy and flexibility of an implementation. Furthermore, this represents a tradeoff between the architectural efficiency and the design time: The higher design effort for ICORE-II results in higher performance, lower area and higher energy-efficiency. However, this design effort can be reduced by better tool support: Due to the microcode programmability of ICORE-II, the current LISA hardware generation capability could not be fully exploited resulting in an increased design time. For future designs it will be possible to model the processor differently in order to overcome this drawback. This approach would have reduced the design time of ICORE-II to about 5 weeks.

It is interesting, that the benchmark runtime of the two applications are similar. This fact makes the library-based ALICE approach attractive for applications, that require medium to high computational performance without tight energy and area constraints.

Despite of the shorter design time of a library-based ASIP design approach, the application-specific optimized processor results in a significantly higher architectural and design efficiency. For applications with tight performance, energy and area constraints, this design approach is clearly superior to the library-based design methodology.

	ICORE-II	ALICE
crit. path	6.8ns	4.9ns
max. area (ND2 equ.) A	34.5k	98.2k
avg. power	47.2mW	261mW
avg. benchmark runtime T	0.32ms	0.58ms
benchmark energy E	15.10μJ	151.4μJ
norm. area	35.1%	100%
norm. power	18.1%	100%
norm. benchmark runtime	55.2%	100%
norm. benchmark energy	9.97%	100%
design time T_{design}	8 weeks (cf. text)	3 weeks[12]
normalized architectural efficiency $1/(ATE)$	100%	1.93%
normalized design efficiency $1/(ATET_{design})$	100%	5.16%

Table 7.5: Comparison Between ICORE-II And ALICE

For higher computational performance requirements, multiple ALICE or multiple ICORE-II instances can be used in order to compute several orthogonal EVD tasks in parallel: This resulting design corresponds to a multi-processor implementation, which has not been further investigated in this thesis.

7.3 Concluding Remarks

The DVB-T A&T case study in Section 7.1 demonstrates the effect of best-practice ASIP design on the computational performance and the energy-efficiency of an implementation. The investigated optimization techniques include the selection of an appropriate ASIP instruction set class, iterative instruction set optimization and further energy optimizations using logic reorganization, clock gating and automatic instruction encoding for energy minimization in the instruction memory. The results of these optimizations show that an energy-efficiency gain of more than one order of magnitude can be achieved.

[12]This design time assumes that ALICE is a predesigned and verified library component with a verified parameterizable HLL compiler.

The case study in Section 7.2 compares a parameterizable processor core with a fully optimized ASIP targeting eigenvalue/eigenvector decomposition of hermitian matrices. The results clearly indicate, that the optimized ASIP is far superior to the parameterizable core concerning energy-efficiency and implementation area. On the other hand, the design time of the parameterizable processor with coprocessor outperforms the fully hand-optimized ASIP. This decrease in design time can be achieved, because the parameterizable processor can be taken from a processor library together with a suitable parameterizable compiler. Obviously, the optimum design methodology strongly depends on the constraints of a given application: For design time critical projects the library-based approach is superior to the constructive ASIP design approach. For designs requiring a high energy-efficiency the constructively designed ASIP is a much better choice.

This case study emphasizes the importance of the HDL generation capability of the LISA design environment. In order to exploit these capabilities, the LISA modeling style of complex processor architectures needs to be optimized using design guidelines in analogy to HLL or HDL coding guidelines. With such an optimized design style, the design efficiency of the constructive ASIP design approach can be increased using the LISA processor design environment. Obviously, this approach can be combined with library-based processor templates, which can be used as a starting point for optimization. This methodology combines the advantages of the constructive with the library-based ASIP design approach.

Chapter 8

Summary

Today's ever-increasing complexities of embedded systems together with tightening time-to-market constraints are the primary drivers for new enabling technologies to enhance the design productivity. State of the art applications require more and more flexibility and functionality of embedded devices, which favors programmable implementations over dedicated hardware. For many handheld appliances like mobile phones and organizers, the battery runtime is almost as important as new functionalities. In previous publications [1] [92] it was demonstrated that high flexibility as well as high performance on the one hand, and high energy-efficiency on the other hand are competing goals. This fact motivates the exploration of new implementation paradigms that enable to trade-off these parameters to optimally satisfy the requirements of an application.

In this thesis, the ability of application-specific instruction set processors (ASIPs) to smoothly trade-off computational performance and flexibility for energy-efficiency is demonstrated. ASIPs are instruction set oriented processors with user-defined instructions, a user-defined data path and, optionally, a more dedicated user-defined accelerator. It is shown that higher performance and increased energy-efficiency can be obtained by exploiting application-specific optimization of the user-defined parts of the ASIP. This specialization removes the upper computational performance bound of traditional fixed processor architectures by introducing the architecture as an additional degree of freedom in the design flow. This enables ASIPs to bridge the performance and energy-efficiency gap between inflexible dedicated hardware and general purpose processors. The quantitative evaluation of a case study shows an ASIP ATE-efficiency[1] that is more than one order of magnitude better than the ATE-efficiency of a general purpose processor.

A major obstacle of ASIP design is the larger design space compared to pure hardware or pure software implementations often resulting in a

[1] This means the equally weighted efficiency for area, time (delay), and energy.

considerably longer design time, which is incompatible with short time-to-market constraints. This issue is the primary motivation of this thesis to identify the time critical tasks in the ASIP design flow and to develop a design methodology with the goal to speed up these tasks.

The contribution of this thesis can be subdivided into the following two tightly related topics:

- enhanced ASIP design flow to obtain a competitive time-to-market (*optimum implementation efficiency*)
- ASIP design optimization for performance and low energy consumption (*optimum architectural efficiency*)

One important challenge of ASIP design is the huge design space, which needs to be explored by the designer in order to obtain an optimum implementation. This thesis proposes an **ASIP design flow** that reduces the design time by using the high level design entry language LISA. LISA has been developed at the Institute for Signal Processing Systems (ISS) together with tools to automate the generation of a complete software design tool chain for a given processor architecture. A LISA description uses a C-based abstraction level concerning the behavior of single LISA operations paired with the concept of concurrency between different operations. This high level description for ASIPs allows design reuse for many product cycles and increases the design productivity.

The design approach proposed in this thesis requires a synthesizable hardware description in the design iteration loop in order to track the impact of high-level decisions like instruction set modifications on important low-level implementation parameters. Examples for these low-level parameters are the critical path, the energy consumption and the silicon area of the ASIP hardware. For this purpose, automatic hardware description generation is needed in order to reduce the time for one design iteration. This thesis contributes essential concepts for this new automatic hardware description generation tool by providing hand-optimized processor cores as case studies and references. Furthermore, critical design decisions for performance and energy consumption are quantitatively identified. Additionally, a concept to ease the development of ASIPs using tool-based automatic instruction encoding is developed. Finally, a methodology for the tedious verification of the final hardware description is presented and a semi-automatic test program generator supporting this approach is described. The proposed design approach is quantitatively evaluated with several case studies, and its efficiency is compared to an alternative library-only-based ASIP design flow without application-specific optimizations. The results of this case study clearly demonstrate that the proposed iterative design approach enables a competitive time-to-market.

Architectural ASIP design optimizations are of paramount importance in order to satisfy the constraints of an application. In this thesis, the design space for ASIP architectures is explicitly defined, thus, providing the basis for any hardware design decision in the ASIP design flow. Moreover, architectural modifications are classified w.r.t. the impact on performance, energy consumption and silicon area in order to provide a sound basis for these critical design decisions. The rationale of increasing the energy-efficiency using typical ASIP specializations is explained in detail. Additional ASIP-typical energy optimizations are implemented and integrated in the design methodology, which substantially improve the total energy-efficiency by about one order of magnitude.

Certain other topics related to ASIP design methodology are beyond the scope of this thesis and are interesting for further research. The development of additional tools to automate certain ever-recurring ASIP optimizations like instruction set specialization is critical to further reduce the design time for ASIPs. Furthermore, the quality of an automatically generated hardware description and of automatically generated compilers are essential factors for the efficiency of the final implementation and the success of this high level design approach.

Appendix A

ASIP Development Using LISA 2.0

In this appendix, the language LISA 2.0, which is the basis of a unified approach for all phases of Application Specific Instruction Set Processor (ASIP) design, is presented. These phases include architecture exploration, architecture implementation, software tools design and architecture integration. The work presented is the result of research at the Institute for Integrated Signal Processing Systems (ISS), Aachen University of Technology, headed by Prof. Heinrich Meyr, Prof. Gerd Ascheid and Prof. Rainer Leupers. This appendix reflects the current research status (October 2003), while major research work is ongoing in the field of compiler generation, Register Transfer Level (RTL) processor synthesis and system integration. The technology developed is commercialized by CoWare Inc. [61].

A.1 The LISA 2.0 Language

The open language LISA 2.0 [110][109] is aimed at the formalized description of programmable architectures, their peripherals and interfaces. It was developed to close the gap between purely structure-oriented languages (VHDL, Verilog) and instruction set languages for architecture exploration purposes. LISA provides a high flexibility to describe the instruction set of various processor types, such as SIMD, MIMD and VLIW-type architectures. Processors with complex pipelines or multi-threading can easily be modelled, too.
Furthermore, LISA models may cover a wide range of abstraction levels. This comprises all levels starting at a pure functional abstraction modelling the data path of the architecture, to a model including the pipeline and functional units. In the domain of timing, the abstraction can go from an instruction-accurate level to cycle-accurate or even

phase-accurate level. A working set of software tools can be generated from all levels of abstraction. Moreover, cycle-accurate models can be used to generate a RTL representation of the architecture.
LISA architecture descriptions are composed of two main components: the resource definition in the so called `RESOURCE` section and the LISA operation tree consisting of several LISA operations. The `RESOURCE` section is a unique place to declare the resources of the architecture such as memories, buses, registers, pipelines and pins. The amount of information given in the `RESOURCE` section depends on the level of abstraction the model is dedicated for. For example, a pipeline is not specified in an instruction-accurate model.

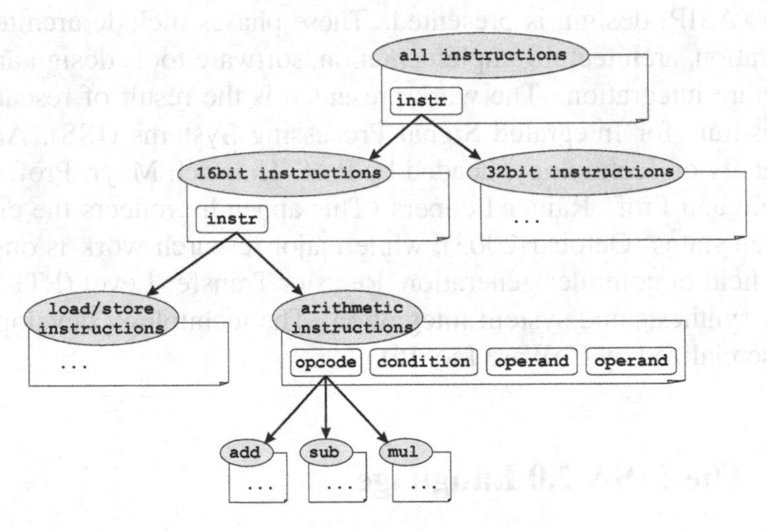

Figure A.1: Extract of the LISA operation tree

A LISA operation consists of various information and is the atomic element of the LISA operation tree. There are two main aspects which must be described explicitly by the LISA operation: the behavior and the instruction set. The behavior is described in the so called `BEHAVIOR` and `EXPRESSION` sections. While the `EXPRESSION` section simply returns a particular value, e.g. a register content, the `BEHAVIOR` section contains the state transition functions of the processor architecture. This state transition is described by writing C code. An instruction set is defined by its assembly syntax and its binary representation. These two pieces of information are described in the LISA `SYNTAX` and `CODING` section, respectively.

Additionally, a LISA operation may contain an `ACTIVATION` section, which describes the timing of the architecture by defining a chain of LISA operations to be executed.
The LISA operations are organized in a tree-like structure. An example can be seen in Figure A.1. The behavior, coding and syntax of an instruction is distributed over several LISA operations. It starts at the root operation, which contains the basic information for all valid processor instructions. In this example, the separation into 16 bit and 32 bit instructions is a first specialization. Each of those operations contains the relevant information for their instruction type. Accordingly, the operations representing the load/store instructions or arithmetic instructions are further specializations of the instructions. The specialization is the basic principle in developing a LISA model. Moreover, as can also be seen in Figure A.1, a LISA operation is not only used to represent a whole instruction but also a *part* of an instruction, such as opcode, operand or special condition field. Thus, developing a LISA model results in creating a LISA operation tree, that unifies the complete description of the behavior, syntax, coding and timing of the target architecture.
The LISA language allows to describe hierarchical models which guarantees modularity and reusability. Architecture models can easily be modified or adopted to new processors, which is the basis of a successful and fast design space exploration.

A.2 Design Space Exploration

The key factor of designing ASIPs is an efficient design space exploration phase. The LISA language allows to apply changes to the architecture model quickly as the level of abstraction is higher than RTL. As shown in Figure A.2 a LISA model of the target architecture is used to automatically generate software tools such as C-compiler, assembler, linker and simulator. These software tools are used to profile and modify both architecture and application. This exploration loop is repeated until a sufficient cost/performance ratio is reached.

Although the higher level of abstraction is the basic reason for the success of Architecture Description Languages (ADLs), the link to the physical parameters such as chip area, power consumption or clock speed gets lost. Ignoring physical parameters in the design space exploration phase leads to suboptimal solutions or long redesign cycles. The necessity of combining the high level abstraction and physical parameter evaluation during design space exploration is compelling.

To overcome those limitations, as shown in Figure A.2, a complete hardware model is automatically generated from LISA in order to get a preliminary estimate about the clock speed, area and power consumption. The LISA processor design platform takes the gate-level synthesis results into account during the exploration phase. The LISA model is used to derive a fully synthesizable model on RTL. Compared to other ASIP development approaches the designer is able to perform this synthesis flow without being restricted to fixed RTL components.

If the synthesis results of the generated architecture fulfill the given physical constraints, then the hardware model can even be used for the final architecture implementation. As the datapath is often highly optimized and based on in-house IP, it may be replaced by the designer manually. This is shown in Figure A.2 on the right hand side.

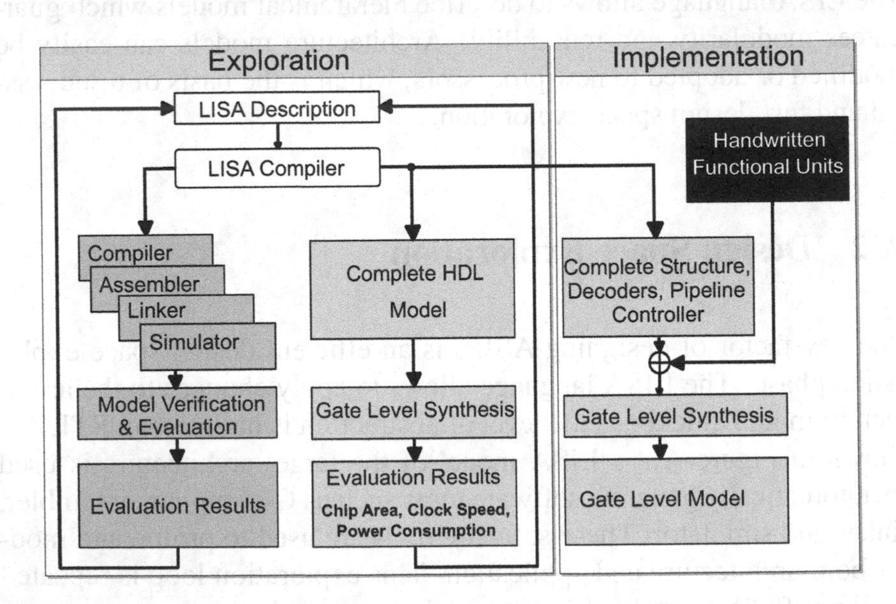

Figure A.2: Exploration and Implementation based on LISA

A.3 Design Implementation

The LISA model is used to derive the complete target architecture in form of a Hardware Description Language (HDL) [217][112]. Languages supported are VHDL, Verilog and SystemC. As described in Section A.2, the generated model is used for design exploration and implementation.

The synthesized architecture consists of several entities. The base entity instantiates one entity which groups all registers, one entity for all memories and another for the complete pipeline. The pipeline entity again consists of entities representing every pipeline stage and intermediate pipeline registers. This entity also contains the automatically generated pipeline controller. The entities representing each pipeline stage instantiates the final level of hierarchy - the entities for the functional units, such as ALUs, Address Generation units, etc. Moreover, the generated decoder is placed inside the pipeline stage entities.

The elements, which constitute the control path, are the instruction decoder and the pipeline controller. The decoder may be distributed over several pipeline stages and sets the control signals to the functional units, to initiate the execution of those. They also steers the pipeline controller. The pipeline controller gathers several information, such as signals from the decoder or the status of the processor, and sends appropriate *flush* and *stall* signals to the pipeline registers.
The decoder generation requires various information about the target architecture. Both, the `RESOURCE` section as well as the LISA operations are used to derive the decoder and control-path. In fact, detailed information about the instruction coding (`CODING` section) and the timing (`ACTIVATION` section) is extracted from the LISA operation tree.
A single LISA operation is assigned to a dedicated pipeline stage. The behavior of a software instruction, for example the instruction `add`, is distributed over several different LISA operations as shown in Figure A.3 :

- The operation `decode` is assigned to the `DE` stage. This operation loads the operands from a general purpose register into a pipeline register.

- The operation `addition`, which adds the values in the pipeline registers and writes the result back to another pipeline register. This operation is assigned to the `EX` stage

- The operation `writeback`, which is assigned to the `WB` stage, writing the value from the pipeline register to the general purpose registers.

The operation execution depends on the LISA timing model. As indicated by the arrows on the left side of Figure A.3, LISA operations activate other LISA operations, which are executed according to their spatial delay in the pipeline. These activation sequences are translated to control signals in the HDL model, which are set or reset depending on the instruction coding of the respective LISA operation.
Decoders are generated in each stage, where activation signals start. Thus, the timing of the architecture is reproduced in the HDL model and the designer might influence the resulting hardware directly via the LISA model. In this example, two decoders are generated, one in the `DE` stage and another in the `EX` stage. If the activation sequence is changed in such a manner that the decode operation activates all other LISA operations, only one single decoder in the `DE` pipeline stage will be generated.

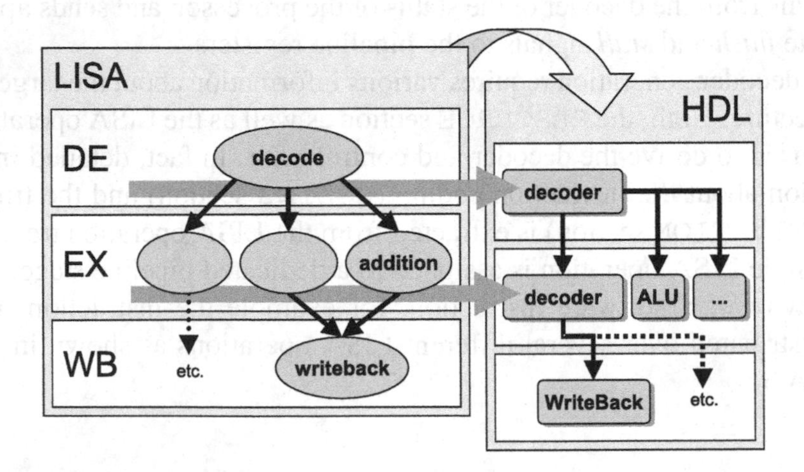

Figure A.3: LISA operation tree and decoder generation

A.4 Software Tools Generation

The software tools generated from the LISA description are able to cope with the requirements of complex application development. From LISA, C-compiler, assembler, linker and simulator are generated. The automatic C-compiler generation is currently one of the major research topics.

A.4.1 Compiler Generation

The shift from assembly to the C programming language for application development is ongoing. This move is driven by the fact that most DSP algorithms are realized using the C language. Considering the various configurations of an ASIP, even during design space exploration, the automatic re-targeting of a compiler is highly desired [267]. For this reason, the automatic generation of a C-compiler came strongly into focus recently.
For retargeting a compiler, the architecture specific back-end of a compiler must be adjusted or rewritten, whereas the architecture independent frontend and most of the optimizations are kept unchanged. Therefore, a retargetable compiler platform is employed which reads in a set of description files generated from the LISA description to build the compiler.
All relevant information for compiler generation are derived from the LISA model. While some information is explicit in the LISA model (e.g. via resource declarations), other relevant information (e.g. concerning instruction scheduling) is only implicit and needs to be extracted by special algorithms. Some further, heavily compiler-specific, information is not at all present in the LISA model, e.g. C type bit widths. Thus, compiler information is automatically extracted from LISA whenever possible, while GUI-based user interaction is employed for other compiler components. The GUI reads the LISA model and presents all relevant machine features (e.g. resources and machine operations) for which interaction is required to the user for further refinement.

The compiler backend basically consists of a register allocator, instruction selector, scheduler and code emitter. Apart from that, the calling conventions and the stacklayout have to be configured. The GUI guides the designer through the specification of the different components:

- Purely numerical parameters (such as C type bit widths, type alignments, minimum addressable memory unit size) are directly captured by means of GUI tables.

- Calling conventions (i.e. how arguments are passed/returned to/from functions) are also captured with GUI tables.

- For the supported stack layout, the designer has to specify the stack pointer and the frame pointer whereas other configuration items can be simply selected/deselected.
- Retargeting the register allocator is reduced to the selection of allocatable registers out of the set of all available registers in the LISA model.
- The scheduler and the code emitter is generated fully automatically [270][268][269].

- The code selector rules are specified by means of a convenient drag-and-drop mechanism: The user can compose these rules from the compiler operators (e.g. addition). Like in most compilers, these mapping rules are the basis for the tree pattern matching based code selector. The link between mapping rules and their arguments on the one hand and LISA operations and their operands on the other hand is made via drag-and-drop in the GUI.

Once everything is specified the designer can finally build the compiler within minutes.

A.4.2 Assembler and Linker Generation

The generated assembler [111] processes the assembly application and produces object code for the target architecture. An automatically generated assembler is required, as the modelled architecture consists of a specialized instruction set. Certainly, the common assembler features are also supported in the generated assembler. For example, many GNU assembler directives are supported. A comfortable macro assembler exists to provide more flexibility to the designer.

The different pieces of object code are linked by the automatically generated linker. With respect to the modelled memory configuration the object code is used to create the final executable. Various configuration possibilities are provided to steer the linking process.

A.4.3 Simulator Generation

The generated simulator is separated into backend and frontend. The debugger frontend and profiler is shown in Figure A.4. It supports application debugging, architecture profiling and application profiling capabilities. The screenshot shows some features such as disassembly view (1) including loop and execution profiling (2), LISA operation execution profiling (3), memory profiling (4) and LISA operation code coverage (5). Also, the content of memories (6), resources (7) and registers (8) can be viewed and modified. Thus, the designer is able to easily debug both the processor model and application. Additionally, the necessary profiling information for design space exploration is provided.

The performance of the simulator is strongly dependent on the abstraction level of the underlying LISA model and the memory model. Figure A.5 shows the ranges of simulation speed achieved by the simulators generated from LISA. The results were achieved by using a 2000 MHz Athlon PC, 768 MB RAM running the Red Hat Linux operating system. The simulation speed of a LISA model, written on a high level of abstraction, both in the domain of timing and architectural features, reaches up to 15 Million Instructions Per Second (MIPS). After increasing the model accuracy, by changing the memory to a complex memory subsystem, the simulation speed drops to 8 MIPS. Changing the core model to a pipelined and thus cycle-accurate version without touching the memory model, decreases the simulation speed by 10 MIPS. Finally, simulating a very detailed model close or equal to the real hardware behavior, the simulator still achieves a speed of about 0,5 MIPS.

The simulator backend, includes a well defined Application Programming Interface (API), which can be easily used to connect to any other

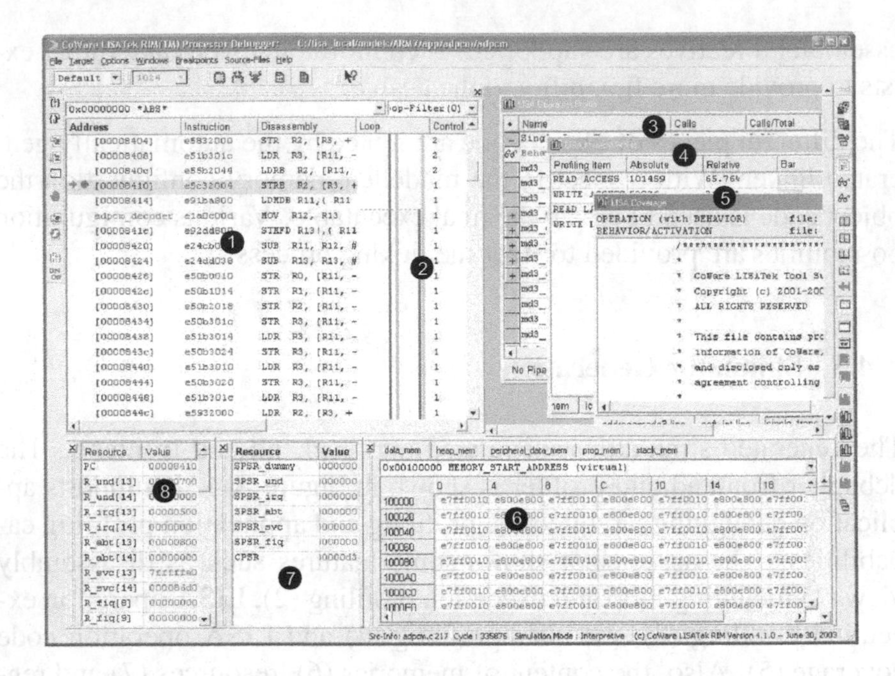

Figure A.4: The simulator and debugger frontend

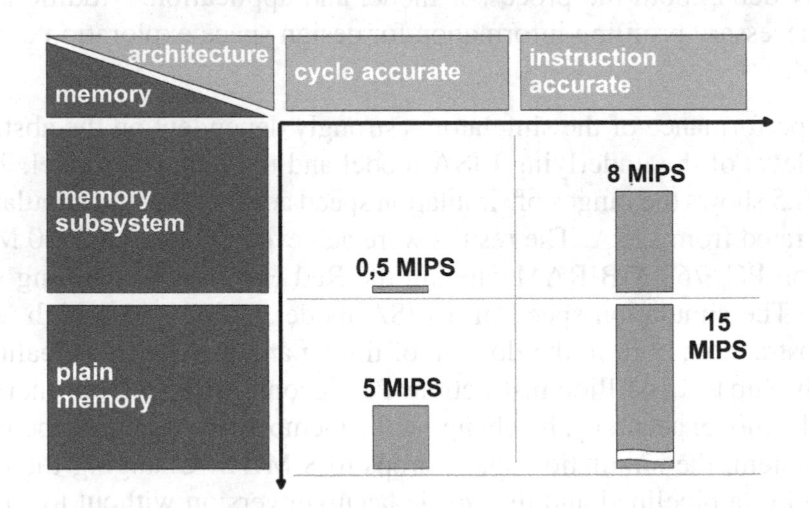

Figure A.5: Achieved simulation speed

simulator frontend. Various simulation techniques [35] are supported, such as compiled simulation, interpretive simulation and Just-In-Time

Cache Compiled Simulation (JIT-CCS) [193]. These mechanisms are briefly described below.

A.4.3.1 Interpretive Simulation

The interpretive simulation technique is a software implementation of the underlying decoder of the architecture. For this reason the interpretive simulation is considered to be a virtual machine performing the same operations as the hardware does: fetch, decode and execute the instruction. All simulation steps are performed at runtime, which provides the highest possible flexibility. However, the straight-forward mapping of the hardware behavior into a software simulator is the major disadvantage of the interpretive simulation technique. Compared to the decoding of the real instructions in hardware, the control flow requires an significant amount of time in software.

A.4.3.2 Compiled Simulation

The compiled simulation uses the locality of code in order to speed up the execution time of the simulation compared to the interpretive simulation technique. The task of fetching and decoding an instruction is performed once before simulation run. The decoding results are stored and used later on during simulation. Execution time is saved as, during the following executions of the same instruction, the fetch and decode steps do not need to be repeated. Thus, the compiled simulation requires the program memory content to be fixed before simulation runtime. Various scenarios are unsupported by the compiled simulation technique, such as system simulations with external and thus unknown memory content and operating systems with changing program memory content. Additionally, large applications, which require a huge amount of memory on the target host, are hard to support.

A.4.3.3 Just-In-Time Cache Compiled Simulation (JIT-CCS)

The objective of the JIT-CCS is to combine the advantages of both interpretive and compiled simulation. This new technique provides the

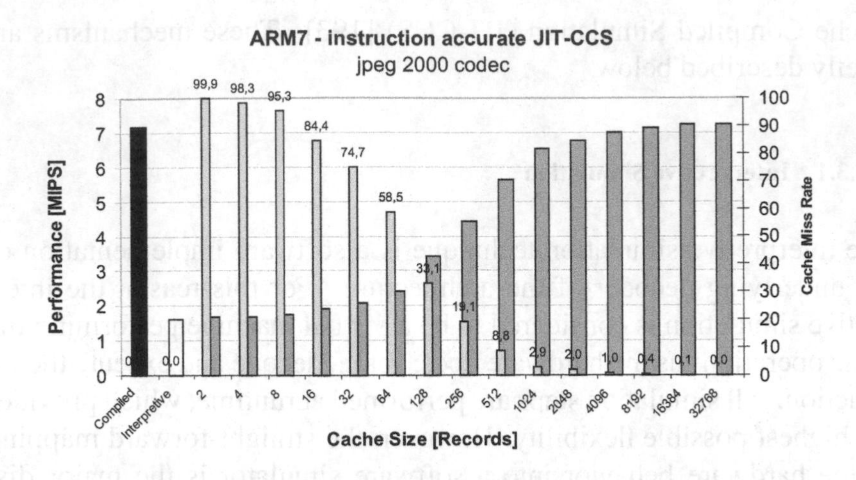

Figure A.6: Performance of the just-in-time cache compiled simulation

full flexibility of the interpretive simulation while reaching the performance of the compiled simulation. The underlying principle is to perform the compilation, in fact the decoding process, *just-in-time* at simulation runtime. Because of that, full flexibility is provided. Moreover, the decoding results are stored in a cache. In every subsequent simulation step the cache is searched for already existing decoding results. Due to the locality of code in typical applications, the simulation speed can be improved using the JIT-CCS. The cache size used in the JIT-CCS is variable and can be changed in a range from 1 - 32768 lines, where a line corresponds to one decoded instruction. The maximum amount of cache lines corresponds to a memory consumption of less than 16 MB on the simulator host. Compared to the traditional compiled simulation technique, where the complete application is translated before simulation time, this memory consumption is negligible.

Figure A.6 illustrates the performance of the cache compiled simulation depending on the cache size. The results were achieved by using an 1200 MHz Athlon PC, 768 MB RAM running the Microsoft Windows 2000 operating system. A cache size of one line means, that the Just-In-Time cache compiled simulation essentially performs the same way as the interpretive simulation. Every instruction is decoded and simulated again, without using the advantage of code locality. With a rising number of cache lines the simulation performance (wide bars) comes closer to the performance of a compiled simulation and even reaches that performance. The simulation speed increases with decreasing cache miss rate (narrow bars). As can be seen in the figure, the performance of the compiled simulation can be reached with a relatively small number of cache size and thus less memory consumption on the host machine. Moreover, this memory consumption on the target host is constant relative to the application size.

A.5 System Integration

Today, typical single chip electronic system implementations combine a mixture of DSPs, micro-controllers, ASICs and memories. In future, the number of programmable units in a System-on-Chip (SoC) design will even increase. To handle the enormous complexity, system level simulation is absolutely necessary for both performance evaluation as well as verification in system context. The earlier in the flow design errors or lack of performance are detected, the less the costs for redesign cycles get. The automatically generated LISA processor simulators can be integrated into various System Simulation Environments, such as CoWare ConvergenSC[61] or SYNOPSYS CoCentric System Studio[234]. Thus, modules provided by different design teams or even third parties can be combined easily.

The communication of the LISA processors with their system environment can be modelled on different levels of abstraction. First, LISA pin resources can directly be mapped to the SoC environment for pin accurate co-simulation. Alternatively, the LISA bus interface allows modelling the SoC communication on a higher abstraction level, i.e., Transaction Level Modeling (TLM) [239]. By that, accesses to buses and memories external to the respective processor core are efficiently mapped to the communication primitives applied in the SoC simulation world.

For user friendly debugging and online profiling of the embedded SW and its platform, the user always has the possibility of getting the full SW centric view of an arbitrary SW block[274] at simulation runtime. This is done just by dynamically connecting a HUB debugger GUI to the processor of interest. Thus, only one or two debugger GUIs are sufficient even to debug complex multiprocessor systems. All other SW blocks, that are currently not considered, still simulate at maximum speed. The remote debugger frontend instance offers all observability and controllability features for multiprocessor simulation as known from standalone processor simulation. Even resources external to a processor module but mapped into its address space like peripheral registers and external memories can be visualized and modified by the multiprocessor debugger GUI. The SW developer can dynamically connect to relevant processors, set break/watch points in respective code segments, disconnect from simulation and automatically re-connect when a breakpoint is hit.

A.6 Summary

This appendix presents the development of Application Specific Instruction Set Processors (ASIPs) based on the architecture description language LISA. This includes the design exploration, implementation, software tools design and system integration. The LISA model of the target architecture is used to automatically generate the software tools: C-compiler, assembler, linker and simulator from the same LISA model. Given a cycle-accurate LISA model, even the complete hardware model can be derived both for exploration and implementation purpose. The generated software tools are powerful enough to be used in complex application design as powerful assembler, macro assembler and different simulation techniques are provided. The highly flexible system integration allows to connect to any co-simulation environment or customer specific environments. The current research topics are focusing on the field of compiler generation, RTL processor synthesis and system integration.

Appendix B

Computational Kernels

In this chapter, the computational kernels that have been used for illustration purposes in Chapter 5 and as benchmarks in Chapter 7 are described.

B.1 The CORDIC Algorithm

The CORDIC algorithm[1] for the vectoring and rotate mode was first described by Volder [266]. In the vectoring mode the magnitude and the angle of a given vector are computed, whereas in the rotate mode a given vector is rotated by a given angle.

The CORDIC algorithm uses iterative computations according to the following equations:

$$X_{i+1} = X_i \mp 2^{-i} Y_i \quad \text{(B.1)}$$

$$Y_{i+1} = Y_i \pm 2^{-i} X_i \quad \text{(B.2)}$$

$$\alpha_{i+1} = \alpha_i \mp tan^{-1}(2^{-i}) \quad \text{(B.3)}$$

These iterations are valid for $X_i > 0$, otherwise, the vector with a negative X_i has to be rotated by 180°. The pathological case of a zero vector has to be treated as an exception.

After N iterations the magnitude of the resulting vector (X_N, Y_N) has been incremented compared to the start vector (X_0, Y_0) by a factor K_N which can be (pre-)computed using the following equation:

$$K_N = \prod_{i=0}^{i=N-1} \sqrt{1 + 2^{-2i}} \quad \text{(B.4)}$$

[1] *CORDIC* stands for COordinate Rotation DIgital Computer.

If the length of the resulting vector in the rotate mode has to be preserved, the scaling factor K_N^{-1} needs to be applied.

The strategy for the choice of the signs in equations B.1 to B.3 has to be selected according to the operating mode.

- rotate mode: in order to obtain $\alpha_N \rightarrow 0$ the $|\alpha_{i+1}|$ has to be smaller than the previous $|\alpha_i|$, therefore, the upper sign is chosen, if $\alpha_i > 0$, else the lower sign.

- vectoring mode: in order to obtain $Y_N \rightarrow 0$ the $|Y_{i+1}|$ has to be smaller than the previous $|Y_i|$, therefore, the lower sign is chosen, if $Y_i > 0$, else the upper sign.

Listings B.1 and B.2 are C code implementations of the CORDIC both for the vectoring and the rotate mode.

```
void cordic_vect(long *x,long y,long *z, const long N) {
  long x_next, y_next, z_next, delta, i, flag;
  double K;
  *z=0; K=1.0;
  if (*x >= 0 )
    flag = 1;
  else {
    *x=-(*x); y=-y; flag = -1;
  }
  for(i=0; i<=N; i++) {       /* "Vectoring" mode y->0 */
    if (y>=0) delta = -1; else if (y<0) delta = 1;
    x_next = (*x)-delta*(y>>i);
    y_next = y+delta*((*x)>>i);
    z_next = (long) (*z-delta*(long) ((1<<N)
             *atan(1.0 / ((float)(1<<i)))+0.5));
    *x = x_next; y  = y_next; *z = z_next;
    K=K*1.0/sqrt(1+pow(2.0, -2.0*i));
  }
  if(flag == -1)
    if (*z < 0) (*z)+=(long) ((1<<N)*M_PI);
    else (*z)-=(long) ((1<<N)*M_PI);
  *x=(long) (K*(*x));    /* scaling */
}
```

Listing B.1: CORDIC Implementation for Vectoring Mode

```
void cordic_rot(long *x,long *y,long z, const long N) {

  long x_next, y_next, z_next, delta, i, flag;
  double K;

  K=1.0;
  if (z<(long)(-(1<<N)*0.5*M_PI)) {
    flag = 1;
    z+=(long) ((1<<N)*M_PI);
  }
  else if(z>(long)((+(1<<N)*0.5*M_PI))) {
    flag = 1;
    z-=(long) ((1<<N)*M_PI);
  }
  else
    flag = 0;

  if ((z<-(1<<N)*0.5*M_PI) || (z>(1<<N)*0.5*M_PI)) {
    printf("\n\nError in CORDIC subroutine: z out of range!\n");
    printf("z = %ld\n", z);
    printf("Bounds: %ld %ld\n", (long) (-(1<<N)*0.5*M_PI),
                          (long ) ((1<<N)*0.5*M_PI));
    exit(1);
  }

  for(i=0; i<=N; i++) {      /* "rotate" mode z->0 */

    if (z>=0)
      delta = 1;
    else if (z<0)
      delta = -1;

    x_next = *x-delta*(*y>>i);
    y_next = *y+delta*(*x>>i);
    z_next = (long) (z-delta*(long) ((1<<N)
                          *atan(1.0 / ((float)(1<<i)))+0.5));
    *x = x_next;
    *y = y_next;
    z = z_next;
    K=K*1.0/sqrt(1+pow(2.0, -2.0*i));
  }

  if (flag == 1) {
    *x=-(*x);
    *y=-(*y);
  }

  *x=(long) (K*(*x));   /* scaling */
  *y=(long) (K*(*y));
}
```

Listing B.2: CORDIC Implementation for Rotate Mode

B.2 FIR Filter

FIR filters are important DSP kernels for a variety of applications. Many commercial DSPs using multiply-accumulate units are optimized for these FIR kernels. The following equation defines the behavior of an M-tap-FIR filter:

$$y(n) = \sum_{k=0}^{k=M-1} h_k x(n-k) \tag{B.5}$$

where $x(m)$ is the input, h_k are the coefficients and $y(n)$ is the output of the filter. In Listing B.3 a C code implementation for the FIR filter is given, which has been taken out of the DSPstone benchmark program suite [285]. This implementation uses explicit memory copy operations in order to obtain the correct delay for the inputs $x(n-k)$. Provided that a processor supports modulo addressing [275], this delay line can also be implemented using a circular buffer in the memory, which approximately halves the number of memory accesses.

B.3 The Fast Fourier Transformation

The fast fourier transformation [57] is an algorithm to compute the discrete fourier transformation (DFT) at reduced computational costs. The 8192 point radix 2 FFT implementation in Listing B.4 uses the decimation in time algorithm [154]. The complex coefficients are partially precomputed, which saves memory bandwidth at the expense of additional arithmetic computations. The function *ReverseBits()* in Listing B.4 is needed to reverse the bit order of an integer for addressing purposes.

B.4 Vector/Matrix Operations

The following vector and vector-matrix operations have been considered:

- dot product: $z = \sum_{i=0}^{i=N-1} x_i y_i$
- matrix-vector multiply: $\mathbf{Z} = \vec{V}^T\mathbf{X}$ and $\mathbf{Z} = \mathbf{X}\vec{V}$
- matrix-matrix multiply: $\mathbf{Z} = \mathbf{XY}$, $\mathbf{Z} = \mathbf{XY}^T$ and $\mathbf{Z} = \mathbf{X}^T\mathbf{Y}$

```
#define STORAGE_CLASS register
#define TYPE  int
#define LENGTH 64

void
pin_down(TYPE * px, TYPE * ph, TYPE y)
{
  STORAGE_CLASS TYPE        i;
  for (i = 1; i <= LENGTH; i++)
    { *px++ = i;
      *ph++ = i;
    }
}

TYPE main()
{
  static TYPE  x[LENGTH];
  static TYPE  h[LENGTH];
  static TYPE  x0 = 100;
  STORAGE_CLASS TYPE i ;
  STORAGE_CLASS TYPE *px, *px2 ;
  STORAGE_CLASS TYPE *ph  ;
  STORAGE_CLASS TYPE y;

  pin_down(x, h, y);
  ph   = &h[LENGTH-1]  ;
  px   = &x[LENGTH-1]  ;
  px2  = &x[LENGTH-2]    ;

//  START_PROFILING ;
  y = 0;
  for (i = 0; i < LENGTH - 1; i++)
    { y += *ph-- * *px ;
      *px-- = *px2-- ;
    }
  y += *ph * *px ;
  *px = x0 ;
//  END_PROFILING ;

  pin_down(x, h, y);
  return ((TYPE) y);
}
```

Listing B.3: Implementation of 64 tap FIR Filter (including Testbench)

- basic element-wise arithmetic: $\mathbf{Z} = X$ op Y and $\vec{z} = \vec{x}$ op $\vec{y}$ where "op" is one of +, -, *
- vector load/store operations (these also support the generation of regular matrices like the identity matrix using programmable stride lengths)
- load/store operation of element in row n and column m in NxM matrix

B.5 Complex EVD using a Jacobi-like Algorithm

According to [134] for the singular value decomposition (SVD) as well as for the eigenvector/eigenvalue decomposition[2] (EVD) there are two algorithms that are widely used: Jacobi-like [128] algorithms and QR-factorization-based algorithms [94]. For this case study, a Jacobi-like algorithm using modified Givens-rotations is used, because of the better numeric stability and precision compared to the QR methods [63].

In order to decompose a hermitian NxN matrix $\mathbf{A} = \mathbf{A_0}$ into the (real) eigenvalues and complex eigenvectors the Givens-rotations are used as follows: After each multiplication according to the following equation, a Pivot element $a_{i,j}$ of the hermitian matrix $\mathbf{A_n}$ is canceled:

$$\mathbf{A}_{n+1} = \mathbf{G}_n^{-1}\mathbf{A}_n\mathbf{G}_n \tag{B.6}$$

The matrix $\mathbf{G}_n$ is the modified Givens-matrix that is computed using a NxN identity matrix into which a 2x2 *pivot submatrix* $\mathbf{Q}_{i,j}$ is embedded at the positions (i,i), (i,j), (j,i) and (j,j) as follows ($i < j$):

$$\mathbf{G}_n = \begin{pmatrix} 1 & & & & & \\ & 1 & & & & \\ & & q_{i,i} & & q_{i,j} & \\ & & & 1 & & \\ & & q_{j,i} & & q_{j,j} & \\ & & & & & 1 \end{pmatrix} \tag{B.7}$$

This 2x2 submatrix $\mathbf{Q}_{i,j}$ is given by

$$\mathbf{Q}_{i,j} = \begin{pmatrix} \cos\varphi & e^{i\alpha}\sin\varphi \\ -e^{-i\alpha}\sin\varphi & \cos\varphi \end{pmatrix} \tag{B.8}$$

where

$$e^{i\alpha} = \frac{a_{i,j}}{|a_{i,j}|} \tag{B.9}$$

and

$$\tan 2\varphi = \frac{2|a_{i,j}|}{a_{j,j} - a_{i,i}}, \qquad \frac{-\pi}{4} \le \varphi \le \frac{\pi}{4} \tag{B.10}$$

[2] The EVD is a special case of the SVD.

Successive Givens-rotations have to be performed for all off-diagonal elements of **A**, which is commonly referred to as one sweep. It is typically necessary to perform several sweeps in order to reach a given precision, because the Givens-rotations set Pivot elements that have already been canceled by a previous rotation back to a value different from zero[3]. After a certain precision is reached after M rotations, which can be monitored by computing the energy of the off-diagonal elements

$$E_{off\_diag} = \sum_{row=1}^{row=N} \sum_{col=row+1}^{col=N} |a_{row,col}|^2 \quad \text{(B.11)}$$

the computed real eigenvalues λ_i are an approximation of the diagonal elements of the matrix $\mathbf{A}_M$.

The associated right and left eigenvectors are given by the orthogonal matrices $\mathbf{EV}_r$ and $\mathbf{EV}_l$ which have normalized rows and columns according to

$$\mathbf{EV}_r = \prod_{i=1}^{M} \mathbf{G}_i \quad \text{(B.12)}$$

and

$$\mathbf{EV}_l = \prod_{i=1}^{M} \mathbf{G}_n^{-1} \quad \text{(B.13)}$$

The computations in Equation B.9 and Equation B.10 can obviously be implemented with a CORDIC processor, which enables to compute the phase and magnitude of a vector as well as the sin() and cos() functions of a given angle.

[3]However, the magnitude of this value is smaller than the magnitude of the value that has just been canceled. Therefore, the algorithm converges to a diagonal matrix.

```
#define DP 10 // decimal point of fixed point numbers
void fft_ll (
    int        InverseTransform, long   *RealIn, long   *ImagIn,
    long  *RealOut, long   *ImagOut )
{
    unsigned i, j, k, n;
    unsigned BlockSize, BlockEnd;
    constant unsigned  NumSamples = 8192; /* FFT-length */
    constant unsigned NumBits=13; /* No. of bits to store indices */
    NumBits = 13;
    long tr, ti;      /* temp real, temp imaginary */
    //    delta_angle;
    long sm2_arr[13] = {0, 0, -1023, -724, -391, -199, -100, -50,
              -25, -12, -6, -3, -1};
    long sm2, sm1_arr[13] = {0, -1023, -724, -391, -199, -100, -50, -25,
                   -12, -6, -3, -1, 0};
    long sm1, cm2_arr[13] = {1023, -1023, 0, 724, 946, 1004, 1019, 1022,
                   1023, 1023, 1023, 1023, 1023};
    long cm2, cm1_arr[13] = {-1023, 0, 724, 946, 1004, 1019, 1022, 1023,
                   1023, 1023, 1023, 1023, 1023};
    long cm1, w, ar[3], ai[3], tmp;
    int loopcnt;
    for ( i=0; i < NumSamples; i++ ) {
        j= ReverseBits ( i, NumBits );
        RealOut[j] = RealIn[i]<<DP;
        ImagOut[j] = (ImagIn == NULL) ? 0 : ImagIn[i]<<DP;
    }
    //  START_PROFILING ;
    loopcnt=0; BlockEnd = 1;
    for ( BlockSize = 2; BlockSize <= NumSamples; BlockSize <<= 1 )
    {
        sm1 = sm1_arr[loopcnt]; sm2 = sm2_arr[loopcnt];
        cm1 = cm1_arr[loopcnt]; cm2 = cm2_arr[loopcnt];
        w = (2 * cm1); loopcnt++;
        for ( i=0; i < NumSamples; i += BlockSize ) {
            ar[2] = cm2; ar[1] = cm1;
            ai[2] = sm2; ai[1] = sm1;
            for ( j=i, n=0; n < BlockEnd; j++, n++ ){
                ar[0]=(w*ar[1])>>DP-ar[2];ar[2]=ar[1];ar[1]=ar[0];
                ai[0](w*ai[1])>>DP-ai[2];ai[2]=ai[1];ai[1]=ai[0];
                k = j + BlockEnd;
                tr = (ar[0]*RealOut[k])>>DP - (ai[0]*ImagOut[k])>>DP;
                ti = (ar[0]*ImagOut[k])>>DP + (ai[0]*RealOut[k])>>DP;
                RealOut[k]=(RealOut[j]-tr); ImagOut[k]=(ImagOut[j]-ti);
                RealOut[j]=(RealOut[j]+tr); ImagOut[j]=(ImagOut[j]+ti);
            }
        }
        BlockEnd = BlockSize;
    }
unsigned ReverseBits ( unsigned index, unsigned NumBits )
{
    unsigned i, rev;
    for ( i=rev=0; i < NumBits; i++ ) {
        rev = (rev << 1) | (index & 1);
        index >>= 1;
    }
    return rev;
} //  END_PROFILING ;
}
```

Listing B.4: Implementation of an 8192 point FFT

Appendix C

ICORE Instruction Set Architecture

This chapter is organized as follows: First of all, the ICORE processor pipeline organization is described and an overview of the important processor resources is given. Furthermore, the processor instructions as well as exceptions to the orthogonal instruction execution model are discussed A description of the memory and I/O organization as well as the ICORE approach to instruction coding concludes this chapter.

C.1 Processor Resources

The visible processor storage entities for the programmer (cf. Figure C.1) are the general purpose register file (8x32bit registers), the address registers (4x9bit), the status register (with less-than and zero flag) and the predicate registers (4x1bit, used as storage bits for conditions). These resources are abbreviated in the following sections according to table C.1 in order to simplify the notation.

The following instruction descriptions use a C-like notation to specify the instruction behavior. Example: AREG=IMM means, that the immediate value "IMM" (which is taken from the instruction word) is loaded into address register "AREG", where "AREG" denotes one of the address registers AR0 to AR3.

C.2 Pipeline Organization

ICORE uses a 3 stage pipeline which is depicted in figure C.2. The first pipeline stage is the stage, where the instruction word is fetched from program memory ("FETCH INSTRUCTION"). The address for

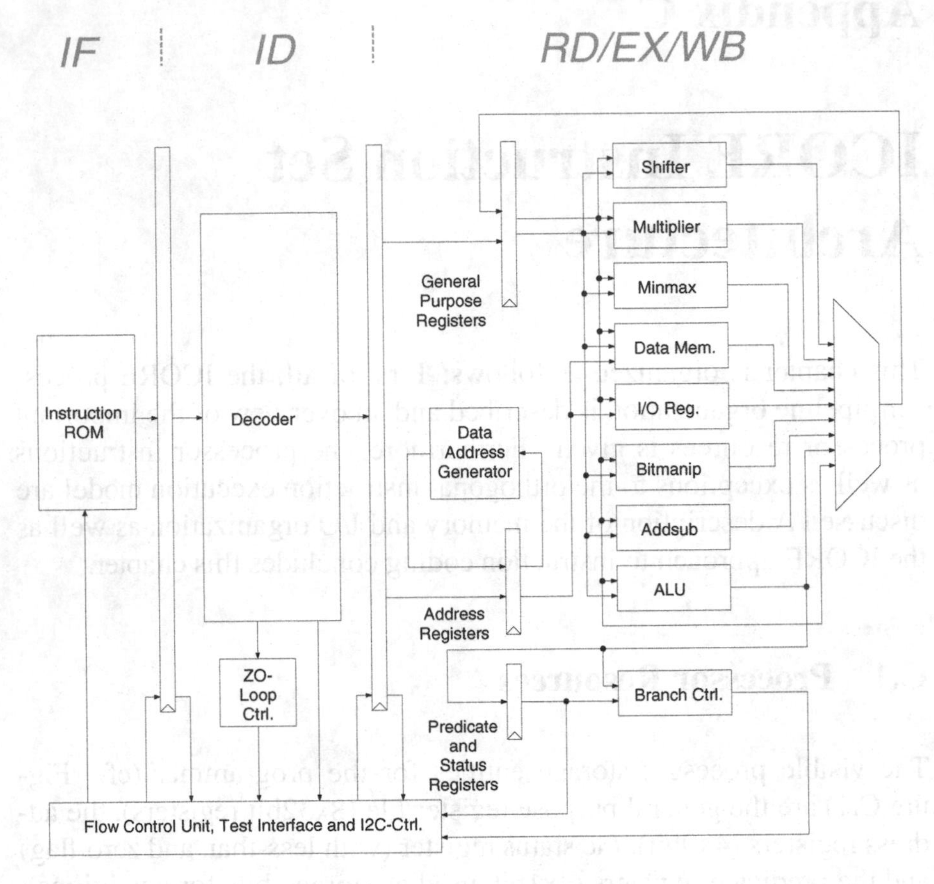

Figure C.1: ICORE Architecture

the program memory is taken from the program counter PC. The program ROM in this stage is clocked by the falling clock edge, whereas all the other registers are clocked by the rising clock edge. Thus, given a certain value for the PC (e.g. PC=0x100), the instruction (in this case the instruction of ROM address 0x100) is stored into the fetch register during the next rising clock edge. This is convenient, because no additional pipeline delay is introduced by the ROM itself. In the next stage ("DECODE INSTRUCTION") the instruction is decoded and internal control signals are generated and stored in the decode register. These signals are propagated to all the functional units of the core and control the behavior of the decoded instruction.

Resource	Abbreviation
REG	one general purpose register (R0-R7)
REGS	source register (which is read)
REGD	destination register (which is written)
AREG	one address register (AR0-AR3) for indirect addressing
IMM	immediate value (constant value encoded in the instruction itself)
FLG	one predicate bit (PR0-PR3)
PC	program counter
STACK	stack for subroutine return address
MEM	data memory
IOPORT	input/output register space
HWL_LOOP_CNT	internal register used as loop counter for the zero-overhead loop

Table C.1: Processor Resources and Abbreviations

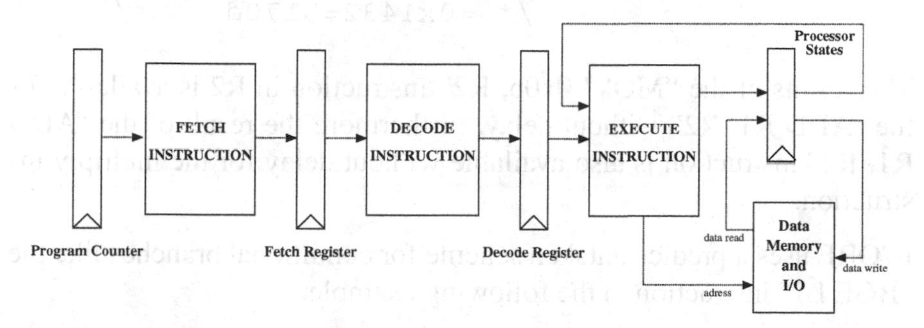

Figure C.2: Abstract ICORE Pipeline Organization

The EXECUTE INSTRUCTION pipeline stage reads and updates the registers storing the processor states (general purpose registers, status register, predicate registers). The most important operations in this pipeline stage are

- register-register operations, reading a general purpose register, executing an operation, and writing back the result to a general purpose register (Examples: multiply or add instructions)
- memory-register operations (load operations) and register-memory operations (store operations) used to transfer data between data

memory and general purpose registers. The same principle is valid for I/O-register (input operations) and register-I/O operations (output operations).

The important point for the DSP programmer is the fact, that the update of the processor state is performed in the same cycle. This means, that no pipeline delay has to be considered for this kind of instructions.

Example 1:

```
MOVI 0x2f, R0     /* R0=0x2f             */
MOVI 0x0a, R1     /* R1=0x0a             */
MOVI 0x0b, R2     /* R1=0x0b             */
ADD  R1,   R2     /* R2=R2+R1            */
MUL  R0,   R2     /* R2=R2*R0=           */
                  /*  (0x0a+0x0b)*0x2f */
                  /*  =0x1432=5170d      */
```

The results of the "MOVI 0x0b, R2" instruction in R2 is available for the "ADD R1, R2" without delay, furthermore the result of the "ADD R1, R2" instruction is also available without delay for the multiply instruction.

ICORE uses a predict-untaken scheme for conditional branches like the "BGE L1" instruction in the following example.

Example 2:

```
   CMP  R1, R2       /* set status reg.   */
   BGE  L1           /* if R2>=R1 then L1 */
   MOVI 0x01, R3     /* R3 = 1            */
   MOVI 0x05, R6     /* R3 = 1            */
   ...
L1:MOVI 0x0, R3      /* R3 = 0            */
```

In this example the program continues without delay with the instructions after the branch ("MOVI 0x01, R3" and "MOVI 0x05, R6"), if the branch is not taken. However, there is a delay after the execution of the branch and the branch target instruction "MOVI 0x0, R3", if the branch is taken. In the case of a taken branch, the pipeline, which has already loaded and decoded both "MOVI" instructions, is flushed (reset to no-operation instructions (NOP)). Thus, the delay between the execution of the taken BGE-instruction and the "MOVI 0x0, R3" is exactly 2 cycles.

In summary, the programmer of ICORE can write functional correct assembly code without having to worry about pipeline delays, because ICORE either has no delays (like in the case of example 1) or the delays slots are hidden from the programmer (like in example 2). This simplifies code development, because the processor behaves according to a straightforward model. An implementation alternative to avoid pipeline flushes after branches would have been to execute the branch delay slots and fill them with useful instructions. The assembly code for this implementation is much less understandable and, frequently, it is impossible to fill the delay slot with a useful instruction other than a NOP. For this reason a very short pipeline for ICORE has been chosen, which implicitly minimizes the branch penalty.

Mnemonic	Description	Behavior
R(AREG,REGD)	Read memory at address AREG and store in register REGD	REGD =MEM(AREG)
RPI(AREG,REGD)	Like "R" with post increment of AREG	REGD =MEM(AREG++)
W(REGS,AREG)	Save register REGS in memory at address AREG	MEM(AREG)= REGS
WPI(REGS,AREG)	Like "W" but with post -increment of AREG	MEM(AREG++) =REGS
IN(AREG,REGD)	Read input ports addressed by AREG	REGD= INPORT(AREG)
INPI(AREG,REGD)	Read input ports addressed by AREG and increment AREG	REGD= INPORT(AREG++)
OUT(REGS,AREG)	Write output port addressed by AREG	OUTPORT(AREG) =REGS
OUTPI(REGS, AREG)	Write output port addressed by AREG and increment AREG	OUTPORT(AREG) =REGS; AREG++

Table C.2: Load/Store Instructions

Mnemonic	Description	Behavior
LAI(CON, AREG)	Load address register immediate	AREG=CON
LAIR0(CON,AREG)	Load address register immediate with displacement in R0	AREG=CON+R0

Table C.3: Address Register Instructions

C.3 Instruction Summary

The ICORE instruction set can be divided into

- 8 load/store instructions (to load (store) data from (to) the data memory or the I/O registers) explained in detail in table C.2
- 28 register-register instructions (performing operations on the general purpose register, using data values from the general purpose registers or the immediate field of the instructions) described in table C.4
- 16 program flow control instructions (like branches, loop instructions and instructions to wait for external events) given in table C.6
- 2 address register load instructions explained in table C.3

The column "Behavior" in the tables C.2, C.4, C.6 and C.3 contains a C-like description of the instruction behavior. The meaning of the operators in this column (e.g. "<<" or "&&") can be looked up in any C language manual. Refer to the appropriate footnote on this page for an explanation of the coding for the "CON"-field of RBIT, WBIT and WBITI.

Mnemonic	Description	Behavior
ABS(REG,FLG)	Abs. value of REG, store sign in pred.	FLG(REG<0):?FLG=1: FLG=0;
ADD(REGS,REGD)	Add two registers, store result in REGD	REGD=REGD+REGS
ADDI(CON, REG)	Add constant value,	REG=REG+CON
ADDSUB0 (REGS,REGD)	Cond. add. or subtraction dep. on the sign of R1	((R1>=0)?(REGD+=REGS) :(REGD-=REGS)
ADDSUB1 (REGS,REGD)	Same as before, but with negated condition	(R1>=0)?(REGD-=REGS) :(REGD+=REGS)
AND(REGS,REGD)	Logical "AND"	REGD = REGD & REGS
ANDI(CON,REGD)	Logical "AND" with constant	REGD = REGD & CON
SAT(CON,REG)	Saturate REG to range from -2^{CON} to $2^{CON}-1$	REG = REG, if in range, else saturation
COR01	Special CORDIC instruction (1. instruction for CORDIC loop)	(R1>=0)?(R3+=R6): (R3-=R6);R5=R7;(R1>=0)? (R2+=R4):(R2-=R4); (HWL_LOOP==-1)? (R7=R2) :(R7= (((R2>>HWL_LOOP) +1)>>1)); R6= MEM [AR0];
COR2	Special CORDIC instruction (2. instruction for CORDIC loop)	(R1<0)?(R1+=R5) :(R1-=R5);HWL_LOOP== -1?(R4=R1):(R4=(((R1>> HWL_LOOP)+1)>>1));
CMP(REGS,REGD)	Compare registers	set_status (REGD-REGS)
CMPI(CON,REG)	Compare with immediate value	set_status(REG-CON)
MOV(REGS,REGD)	Move reg.	REGD=REGS
MOVI(CON,REG)	Move immediate	REG=CON
MULS(REGS,REGD)	Signed mult. of lower 16 bits of REGS and REGD	REGD= ((unsigned_16b) REGS * (unsigned_16b) REGD)
MULU(REGS,REGD)	Unsigned mult. of lower 16 bits of REGS and REGD	REGD= ((unsigned_16b) REGS * (unsigned_16b) REGD)

Table C.4: Register-Register Instructions (Part 1/2)

Mnemonic	Description	Behavior
NEG(REG)	Negate register	REG = -REG
SLA(REGS,REGD)	Arith. shift left (shift count in 5 LSBs of REGS)	REGD=REGD<<REGS_low5
SLAI(CON,REG)	Arith. shift left	REG=REG<<CON_low5
SRA(REGS,REGD)	Arith. shift right	REGD=REGD>>REGS_low5
SRAI(CON,REG)	Arith. shift right	REGD=REG>>CON
SRA1(REGS, REGD)	CORDIC instr.: (shift by nr+1 bits, where nr="current hardware loop cnt.", with rounding)	(HWL_COUNT<= -1)? (REGD=REGS): (REGD= (((REGS>>HWL_COUNT)+1) >>1)
SRAI1(CON,REG)	CORDIC instr. for shift and round	(CON<= -1)?(REG=REG) :(REG=(((REG>>CON)+1) >>1)
SUB(REGS,REGD)	Subtraction	REGD=REGD-REGS
SUBI(CON,REG)	Subtraction of const.	REG=REG-CON
RBIT(CON[4],REG)	Extract bit field in REG	R0=((REG>>CON_right) & ((1<<CON_length)-1))
WBIT(CON[5],REG)	Write bit field in R0	R0=(((R0>>(CON_left+1)) <<(CON_left+1))+((REG& ((1<<(CON_left-CON_right +1))-1))<<CON_right)+ +(R0&((1<<CON_right)-1)))
WBITI(CON[6],REG)	Write constant in bit field of REG	REG=(((REG>>(CON_left+1)) <<(CON_left+1))+((31& CON_value&((1<<(CON_left -CON_right+1))-1)) <<CON_right)+(REG&((1 <<CON_right)-1)))

Table C.5: Register-Register Instructions (Part 2/2)

[4] CON is defined by the relation CON=CON_length*8 +CON_right, where CON_length and CON_right are 3 bit unsigned values

[5] CON is defined by CON=CON_length*8 +CON_right, where CON_length and CON_right are 3 bit unsigned values; CON_left = CON_right + CON_length -1

[6] CON is defined by CON=64*CON_value+8*CON_length+CON_right, where CON_value is a 5 bit field and CON_length and CON_right are 3 bit unsigned values; CON_left = CON_right + CON_length -1

Mnemonic	Description	Behavior
B(CON)	Uncond. rel. branch	PC=PC+CON+1
BSR(CON)	Uncond. rel. branch to subroutine	STACK=PC; PC=PC+CON+1;
BE(CON)	Rel. branch if ”=” cond. in status reg.	If(STATUS.Z==1) PC=PC+CON+1;
BNE(CON)	Rel. branch if ”!=” cond. in status reg.	if(STATUS.Z==0) PC=PC+CON+1;
BLT(CON)	Rel. branch if ”<” cond. in status reg.	if(STATUS.L==1 && STATUS.L==0) PC=PC+CON+1
BLE(CON)	Rel. branch if ”<=” cond. in status register	if(STATUS.L==1 —— STATUS.Z==1) PC=PC+CON+1
BGT(CON)	Rel. branch if ”>” cond. in status register	if(STATUS.L==0 && STATUS.Z==0) PC=PC+CON+1
BGE(CON)	Rel. branch if ”>=” cond. in status register	if(STATUS.L==0 —— STATUS.Z==0) PC=PC+CON+1
BPC(FLG,CON)	Rel. branch if pred. bit FLG is clear	if(FLG==0) PC=PC+CON+1
BPS(FLG,CON)	Rel. branch if pred. bit FLG is set	if(FLG==1) PC=PC+CON+1
END	Enter idle mode	-
RTS	Return from subrout.	PC=STACK
SUSPG	Wait until guard_trig=”1”	-
SUSPP	Wait until ppubus_en=”1”	-
LPCNT(CON,REG)	Init. loop cnt. reg.	lp_start_count=CON; lp_end_count=REG;
LPINI(CON)	Init. loop start and end addr. (CON) and activ. loop processing with next instr.	-

Table C.6: Program Flow Control Instructions

C.4 Exceptions to the Hidden Pipeline Model

There are some restrictions for the programmer due to the internal pipeline of ICORE. Restrictions are only present for the zero-overhead-loop processing, which is implemented by the flow-control unit (cf. Figure C.1).

- the instruction "lpcnt" has to be executed at least 2 instructions before the loop starts, when the internal loop counter is needed within the loop e.g.

```
  LPCNT(1,24);
  MOV(R1,R2);
  LPINI(COR_LOOPEND);
LABEL(COR_LOOPSTART)
  COR01;
  COR2;
LABEL(COR_LOOPEND)
```

  is a legal sequence

- the loop body (in the above example the instructions COR01 and COR2) needs at least 2 instructions. Only one loop instruction in the loop body is not supported.

- branches to the last loop instruction are illegal e.g. in

```
  LPCNT(1,10);
  MOV(R1,R2);
  LPINI(LOOPEND);
LABEL(LOOPSTART)
  RPI(AR0,R0);
  CMP(0,R0);
  BNE(END_BODY);
  MOVI(1,R0);
LABEL(END_BODY)
  WPI(R0,AR1);
LABEL(LOOPEND)
```

  the "BNE (END_BODY);" instruction is illegal

C.5 ICORE Memory Organization and I/O Space

ICORE uses a Harvard architecture, which means that instruction and data memory are separated. An instruction ROM of 2048 words with 20 bits is used to store the program.

The data memory is subdivided into a small (synthesized) 24 word ROM and a 256x32 bit RAM. Table C.7 shows the address mapping of the data memory. The data memory itself contains about 200 temporary states and variables which correspond to the states defined in the DVB-T A&T specification.

Memory	Start Address	End Address
Synth. ROM	0	23
unused	24	255
RAM	256	511

Table C.7: Data Memory Mapping

The I/O-address space of ICORE is separated from the data and the instruction memory and uses about 40 different registers for I/O values.

C.6 Instruction Coding

The *instruction code word* is the representation of operations and operands in the instruction memory. For instance, the word “000100DDDIIIIIIIIIII” is the instruction code word for “MOVI #I, D” where “I” represents an 11-bit immediate value and D is the destination (general purpose register R0-R7). Thus, the operation “MOVI 011100b, R5” moves the signed binary value 011100b to the register R5 and has the machine coding “00010010100000011100”.

In order to simplify the design space exploration, which involves frequent changes of the instruction coding, a tool called ICON has been developed for programming. This tool has been described in Subsection 6.3.1 in this thesis. ICON is an instruction coding generator, a hardware generator for the decoder hardware description, and an assembler. ICON uses the assembler program in a line-oriented input file and automatically generates the instruction coding, the instruction decoder and the machine code (as COFF file for the ROM). The user has the freedom to select the preferred coding scheme for the opcodes and the alignment of operand fields. The remaining degrees of freedom are optimized to minimize the coding width and the power consumption. Figure C.3 shows an example input file (".cri"-file) for ICON, which contains the assembler program in the following format:

- the first line contains a description of the individual fields of the file, starting with the opcode and the individual operands. The second field, for instance, specifies the field "reg0" (first general purpose register) which is encoded as 3bit unsigned value ("3u"), the fifth field specifies the 11-bit signed immediate value
- the second line is a blank line
- the following lines specify the actual program to be implemented starting with the operation e.g. "movi_op" and the appropriate fields e.g. "u3reg0=5" and "s11immediate=28" for the above example ("MOVI 011100b, R5"). "x" values indicate don't care operand values.

```
opcode   |u3r0     |u3r1  |u2ar    |s11imm |u2pr

b_op      x         x      x        6       x
movi_op   0         x      x        0       x
lai_op    x         x      0        32      x
lpc_op    1         x      x        1       x
rpi_op    2         x      1        x       x
cmp_op    4         2      x        x       x
end_op    x         x      x        x       x
```

Figure C.3: Assembler Input File for ICON

Appendix D

Different ICORE Pipeline Organizations

Figure D.1 depicts the different pipeline organizations that have been explored during the design of ICORE. Figure D.1 neglects many details of the implementation including the status and predicate register file, the address generator etc. It rather depicts the part of the data path that is needed for register-register and memory-register instructions. Implementation a) shows a two stage pipeline: the critical path of this implementation is typically the ID/RD/EX/WB stage. Implementation b) reduces this critical path by inserting an additional pipeline stage after the decoder. This also increases the branch penalty by one cycle. Implementation c), finally, uses a 4-stage pipeline with a data forwarding path. This implementation, which is pretty similar to many conventional RISC-processor implementations, needs additional MUXes in the RD/EX stage to implement the forwarding logic. For the ICORE benchmark, implementation b) exposed the best energy-efficiency and was able to meet the given timing constraints.

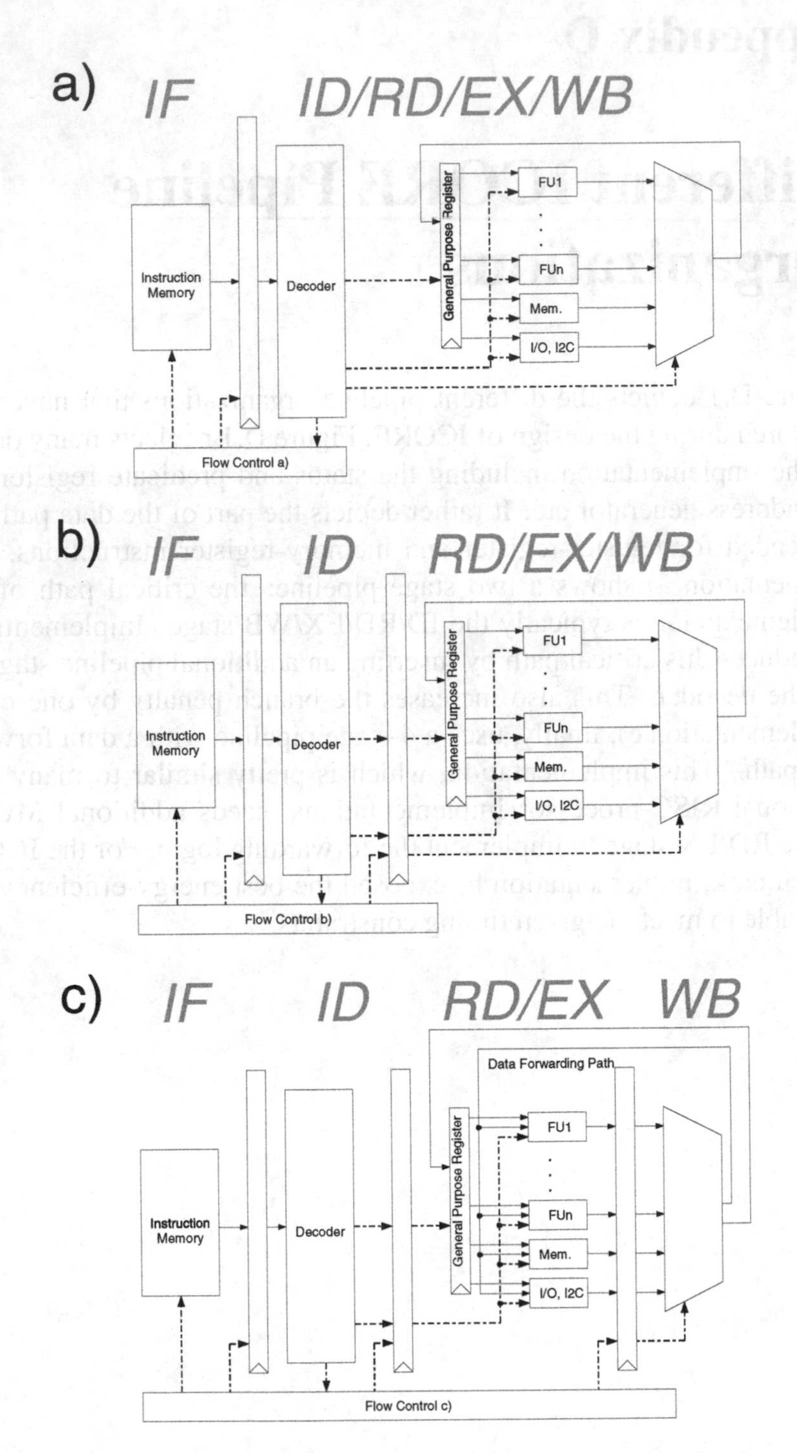

Figure D.1: Different Pipeline Organizations for Design Exploration

Appendix E

ICORE HDL Description Templates

In this chapter generic examples for HDL templates that implement register file instances and functional units are described. These descriptions can also be used and parameterized by an automatic HDL generator.

E.1 Generic Register File Entity

The implementation of a processor register like a status or a general purpose register file can be achieved using the template register description in Listing E.2 together with a parameterization package according to Listing E.1. This synthesizable register file can be parameterized to match any register structure that can be described with LISA.

Parameters of this register file template are the bit width of register elements, the number of registers, and the number of individual read and write ports of the register. If two write ports try to write to the same address, this write access conflict is resolved in hardware using a priorization scheme (here, write ports with lower number have higher priority). Optionally, hardware or simulation code to detect this condition can be added (which has been omitted in Listing E.2 due to a lack of space).

Write and read accesses to such a register file can be implemented by connecting the inputs of the write ports to the associated data sources using multiplexers in case of multiple sources. Due to the fact that the read and write ports have access to all of the available internal registers, the hardware generator has to find an optimum assignment of data sources/sinks to register write/read ports. This assignment can be achieved using a balancing scheme, which minimizes the total multiplexer area and delay of the implementation.

```
LIBRARY ieee;
USE ieee.std_logic_1164.ALL;
USE ieee.std_logic_arith.ALL;

PACKAGE xy_reg_defs IS
-- GENERAL PURPOSE REGISTERS
  CONSTANT xy_reg_width          : integer := 32;
  CONSTANT xy_ldn_num_registers : integer := 3;
-- modify the following, if no power of 2 is required
  CONSTANT xy_num_registers      : integer := 2**xy_ldn_num_registers;
-- number of read and write ports of reg. file
  CONSTANT xy_num_read_ports     : integer := 5;
  CONSTANT xy_num_write_ports    : integer := 5;

-- XY REGISTERS INTERFACES
  SUBTYPE xy_register_t is std_logic_vector(xy_reg_width-1 DOWNTO 0);
-- type to select registers
  SUBTYPE reg_nr_t IS unsigned(xy_ldn_num_registers-1 DOWNTO 0);
-- all read ports
  TYPE xy_read_port_array_t IS ARRAY(xy_num_read_ports-1 DOWNTO 0)
    OF xy_register_t;
-- all write ports
  TYPE xy_write_port_array_t IS ARRAY (xy_num_write_ports-1 DOWNTO 0)
    OF xy_register_t;
-- selects if write port is active and which register is written
  TYPE xy_write_port_enable_array_t IS ARRAY (xy_num_write_ports-1
                                          DOWNTO 0) OF std_logic;

  TYPE xy_write_port_nr_array_t IS ARRAY (xy_num_write_ports-1 DOWNTO 0)
    OF reg_nr_t;
-- selects which register is read
  TYPE xy_read_port_nr_array_t IS ARRAY (xy_num_read_ports-1 DOWNTO 0)
    OF reg_nr_t;
-- the register file itself
  TYPE xy_register_array_t IS ARRAY (xy_num_registers-1 DOWNTO 0)
    OF xy_register_t;
END xy_reg_defs;
```

Listing E.1: Package with Definitions for Parameterizable Register File

1 Read Port, 1 Write Port # of Registers	2	4	8	16	32
Area (ND2 equ. gates)	0.61k	1.09k	2.10k	3.99k	7.79k

1 Write Port, 8 Registers # of Read Ports	1	2	3	4	5
Area (ND2 equ. gates)	2.10k	2.53k	2.99k	3.44k	3.90k

2 Read Ports, 8 Registers # of Write Ports	1	2	3	4	5
Area (ND2 equ. gates)	2.53k	3.15k	3.28k	3.79k	3.88k

Table E.1: Area Results for Example Register File Configurations

Table E.1 contains synthesis results for a register file with 32 bit registers and several example configurations for the register number and the number of read and write ports. Compared to the area of a functional unit like a 16x16 bit multiplier, which consumes about 1.9k equivalent gates (target frequency: 200MHz), the register file area can be significant, if many registers are needed. From the power perspective, registers also consume a significant part of the total power, because they are frequently accessed in a typical load/store architecture (cf. Appendix F).

E.2 Generic Bit-Manipulation Unit

This section describes an application-specific functional unit with the purpose to read and write short bit fields within a longer (e.g. 32-bit) word. This so-called bit-manipulation unit is an example for a hand-optimized VHDL description of a functional unit, because this kind of operation is not available in the DesignWare-library of Synopsys [233].

As an example for a bit field write operation refer to Section C.3, where the instructions RBIT, RBITI, WBIT and WBITI are described. For instance in case of a WBIT instruction, the bit-manipulation unit (cf. VHDL-Listings E.5, E.3 and E.4) simply replaces the bits CON_right to $CON\_right + CON\_length - 1$ with the CON_length LSBs of $op1\_in\_nb$.

```
ENTITY xy_reg_file IS
  PORT (
    clk_sysd2,rstq          : IN  std_logic;
    xy_write_port_enable    : IN  xy_write_port_enable_array_t;
    xy_write_port_regnr     : IN  xy_write_port_nr_array_t;
    xy_data_in              : IN  xy_write_port_array_t;
    xy_regnr_read           : IN  xy_read_port_nr_array_t;
    xy_data_out             : OUT xy_read_port_array_t;
    xy_register             : OUT xy_register_array_t);
END xy_reg_file;

ARCHITECTURE rtl OF xy_reg_file IS
  SIGNAL gpreg : xy_register_array_t;
BEGIN
  xy_register <= gpreg;
  gp_register_write : PROCESS(rstq, clk_sysd2)
   VARIABLE enable_bus : std_logic_vector(xy_num_registers-1
                                          DOWNTO 0);
   VARIABLE tmp_gpreg  : xy_register_t;
  BEGIN
    IF (rstq = '0') THEN
      gpreg <= (OTHERS => (OTHERS => '0'));
    ELSIF (clk_sysd2'event AND clk_sysd2 = '1') THEN
      enable_bus := (OTHERS => '0');
      FOR i IN xy_num_write_ports-1 DOWNTO 0 LOOP
        IF (xy_write_port_enable(i) = '1') THEN
          enable_bus(conv_integer(xy_write_port_regnr(i))) := '1';
        END IF;
      END LOOP;
      FOR i IN xy_num_registers-1 DOWNTO 0 LOOP
        tmp_gpreg := (OTHERS => '0');
        FOR j IN xy_num_write_ports-1 DOWNTO 0 LOOP
          IF (xy_write_port_regnr(j) = i
            AND xy_write_port_enable(j) = '1') THEN
            tmp_gpreg := xy_data_in(j);
          END IF;
        END LOOP;
        IF (enable_bus(i) = '1') THEN
          gpreg(i) <= tmp_gpreg;
        ELSE
          gpreg(i) <= gpreg(i);
        END IF;
      END LOOP;
    END IF;
  END PROCESS;

  gp_register_read : PROCESS(gpreg, xy_regnr_read)
  VARIABLE nr_var : integer range 0 to xy_num_registers-1;
  BEGIN
    xy_data_out <= (OTHERS => (OTHERS => '0'));
    FOR i IN xy_num_read_ports-1 DOWNTO 0 LOOP
      nr_var := conv_integer(xy_regnr_read(i));
      xy_data_out(i) <= gpreg(nr_var);
    END LOOP;
  END PROCESS;
  -- insert write collision detection here
END rtl;
```

Listing E.2: Synthesizable VHDL Description of Parameterizable RF

```
ARCHITECTURE rtl OF ppu_bitmanip IS
  SIGNAL rbit_notwbit : std_logic;
  SIGNAL enable_imm   : std_logic;
  SIGNAL immediate    : immediate_value_t;
  SIGNAL op0_in       : data_path_t;
  SIGNAL op1_in       : data_path_t;
  SUBTYPE mask_t IS unsigned(bitmanip_max_mod_field-1 DOWNTO 0);
  TYPE mask_table_t IS ARRAY (bitmanip_max_mod_field-1 DOWNTO 0) OF mask_t;
  SIGNAL mask_table   : mask_table_t;
  SIGNAL dummy        : std_logic;
BEGIN
  bitmanip_block_proc : PROCESS(enable, rbit_notwbit_nb, enable_imm_nb,
                                immediate_nb, op0_in_nb, op1_in_nb)
  BEGIN
    IF (enable = '0' AND use_blocked_input = 1) THEN
      rbit_notwbit <= '0';
      enable_imm   <= '0';
      immediate    <= (OTHERS => '0');
      op0_in       <= (OTHERS => '0');
      op1_in       <= (OTHERS => '0');
    ELSE
      rbit_notwbit <= rbit_notwbit_nb;
      enable_imm   <= enable_imm_nb;
      immediate    <= immediate_nb;
      op0_in       <= op0_in_nb;
      op1_in       <= op1_in_nb;
    END IF;
  END PROCESS;
  bitmanip_proc : PROCESS(rbit_notwbit, enable_imm, immediate,
                          op0_in, op1_in, mask_table)
    VARIABLE value_to_insert_v: std_logic_vector(bitmanip_max_mod_field-1
                                                  DOWNTO 0);
    VARIABLE masked_value_v    : data_path_t;
    VARIABLE shifted_mask_v    : std_logic_vector(bitmanip_max_affected
                                                   DOWNTO 0);
    VARIABLE shifted_op_v      : data_path_t;
    VARIABLE length_v          : unsigned(bitmanip_len_field_len-1
                                  DOWNTO 0);
    VARIABLE right_pos_v       : unsigned(bitmanip_rpos_field_len-1
                                          DOWNTO 0);
  BEGIN
    length_v := conv_unsigned(immediate(imm_value_width-
                              bitmanip_value_field_le-1 DOWNTO
                              imm_value_width-bitmanip_value_field_len
                             -bitmanip_len_field_len),
                              bitmanip_len_field_len);
    right_pos_v := conv_unsigned(immediate(bitmanip_rpos_field_len-1
                         DOWNTO 0),bitmanip_rpos_field_len);
    shifted_mask_v := conv_std_logic_vector(shl(mask_table(
                      conv_integer(length_v)),
                      right_pos_v), bitmanip_max_affected+1);
    IF (rbit_notwbit = '0') THEN
      IF (enable_imm = '1') THEN           -- wbiti
        value_to_insert_v := conv_std_logic_vector(conv_unsigned(immediate(
        imm_value_width-1 DOWNTO imm_value_width-bitmanip_value_field_len),
        bitmanip_value_field_len),bitmanip_max_mod_field);
      ELSE                                 -- wbit
        value_to_insert_v := op1_in(bitmanip_max_mod_field-1 DOWNTO 0);
      END IF;
```

Listing E.3: Synthesizable VHDL Architecture of Bit-Manipulation Unit 1/2

```
    -- mask significant bits
    masked_value_v := (OTHERS => '0');
    masked_value_v(bitmanip_max_mod_field-1 DOWNTO 0) :=
      value_to_insert_v AND
      conv_std_logic_vector(mask_table(conv_integer(length_v)),
                            bitmanip_max_mod_field);
    -- shift masked_value_v to the left by right_pos_v bits
    masked_value_v := conv_std_logic_vector(shl(UNSIGNED(masked_value_v),
                      right_pos_v), data_path_width);
    -- combine results
    result     <= (op0_in AND (NOT conv_std_logic_vector(
      UNSIGNED(shifted_mask_v),data_path_width)))
                 OR masked_value_v;
  ELSE                                    -- rbit
    -- shift right
    shifted_op_v := conv_std_logic_vector(shr(UNSIGNED(op0_in),
                    right_pos_v), data_path_width);
    -- mask
    result <= shifted_op_v AND
              conv_std_logic_vector(mask_table(
              conv_integer(length_v)),data_path_width);
  END IF;
END PROCESS;

gen_mask_proc : PROCESS(dummy)
BEGIN
  mask_table <= (OTHERS => (OTHERS => '0'));
  FOR j IN bitmanip_max_mod_field-2 DOWNTO 0 LOOP
    FOR i IN j DOWNTO 0 LOOP
      mask_table(j+1)(i) <= '1';
    END LOOP;  -- i
  END LOOP;  -- j
  mask_table(0) <= (OTHERS => '1');
END PROCESS;
END rtl;
```

Listing E.4: Synthesizable VHDL Architecture of Bit-Manipulation Unit 2/2

```
ENTITY ppu_bitmanip IS
  PORT (
    enable          : IN  std_logic;
    rbit_notwbit_nb : IN  std_logic;          -- '0' = wbit instruction
    enable_imm_nb   : IN  std_logic;          -- for wbiti
    immediate_nb    : IN  immediate_value_t;  -- immediate from instruct.
    op0_in_nb       : IN  data_path_t;        -- operand from register
    op1_in_nb       : IN  data_path_t;        -- "value field" for wbit
    result          : OUT data_path_t         -- update for r0
    );
END ppu_bitmanip;
```

Listing E.5: Synthesizable VHDL Entity of Bit-Manipulation Unit

Appendix F

Area, Power and Design Time for ICORE

Figure F.1 depicts the area breakdown of the ICORE design module, which contains the processor core, instruction and data memory as well as an I/O processor. The total area of this design module is about 0.65 sq. mm.

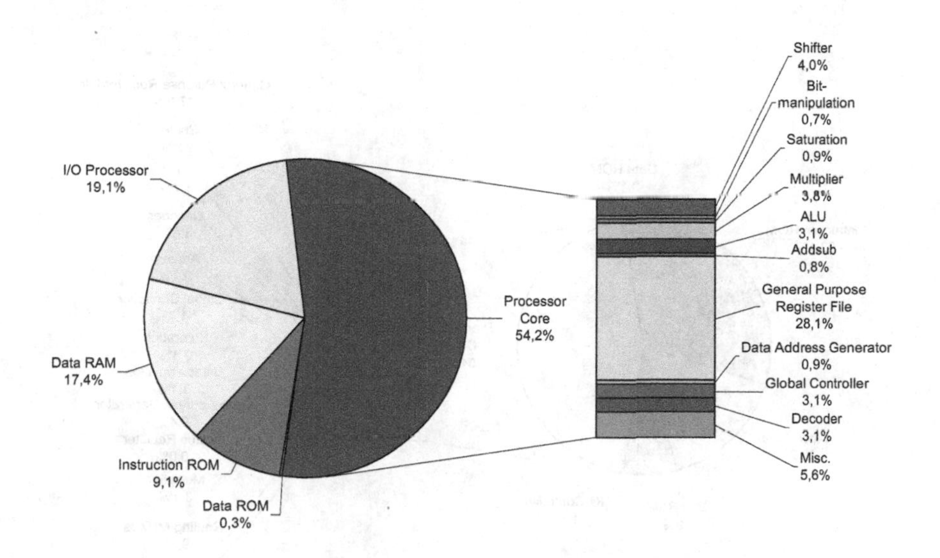

Figure F.1: ICORE Area Breakdown

A significant amount of the area is consumed by design entities that require a large amount of area-intensive D-Flipflops like the general purpose register and the I/O processor. The functional units of the processor consume less than 15% of the total area.

In Figure F.2 the power consumption of the ICORE module for all the design entities as well as the clock tree is shown. The total power of the complete design module is 18.8mW.

Compared to other processor-based systems, the power consumption in the clock tree is significantly lower, due to the extensive use of clock gating. This is especially efficient, because the switching activity of many register intensive design entities is low: For instance, a large part of the I/O controller is used in order to store mainly static information that configures the operating mode of the design. Moreover, it has to be mentioned that the power consumption of the instruction memory in Figure F.2 is measured after the optimization process that has been described in Section 6.3.1. The power in the instruction ROM is still significant despite of this optimization, because the memory is accessed in each clock cycle.

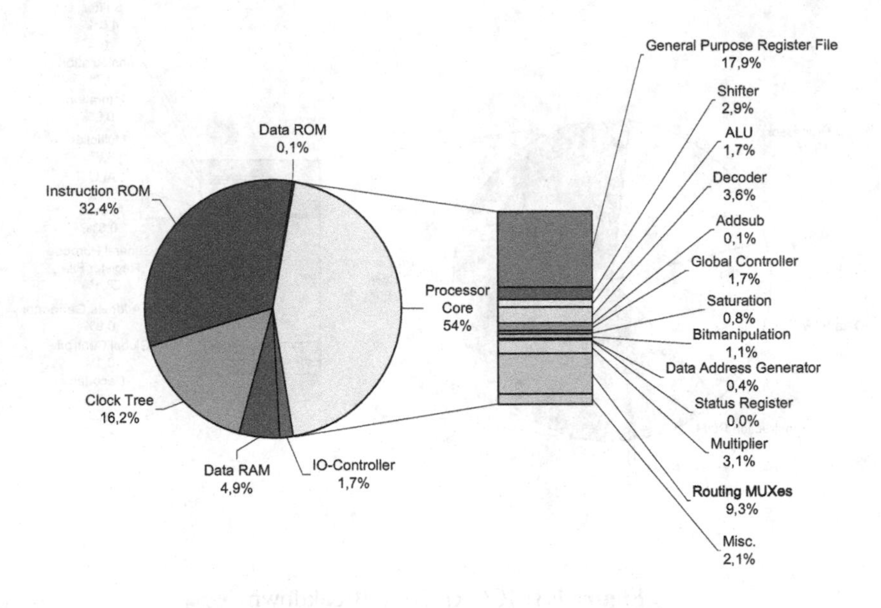

Figure F.2: ICORE Power Breakdown for a Typical Operation Scenario

The total design time of ICORE was 10.5 man months. Figure F.3 shows the percentage of the design time for the different design tasks. The time for the design space exploration is included in the assembly programming and HW description design tasks. The time for all the verification tasks in this case study is significant with about 45% of the total design time. An optimizing (and functionally correct) HLL compiler would have considerably decreased the time for assembly programming and SW verification.

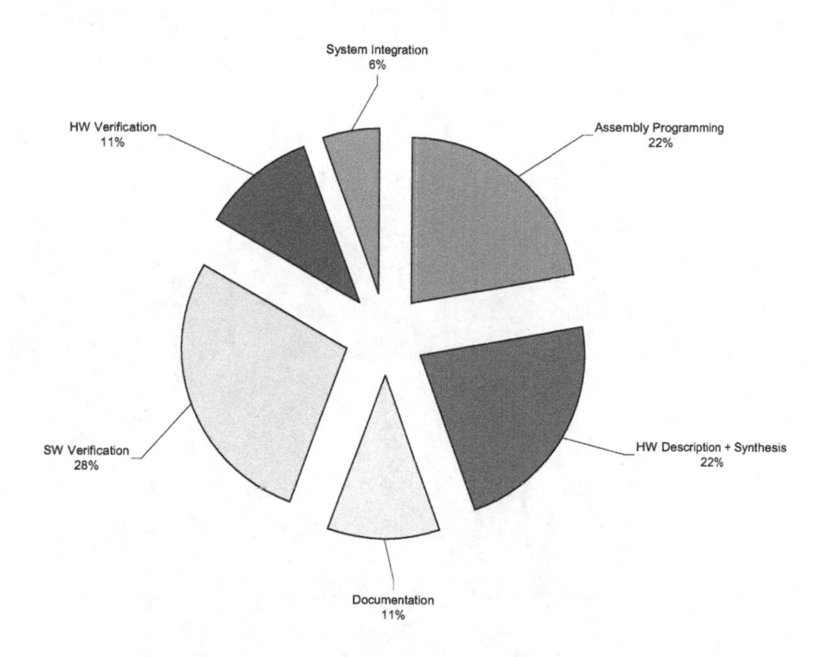

Figure F.3: Distribution of ICORE Designtime

Appendix G

Acronyms

A	silicon area
AD/DA	analog digital / digital analog converter
ALICE	here: name of a parameterizable processor architecture with compiler support
ALU	arithmetic logic unit
API	application programming interface
ASI	application specific instruction
ASIC	application-specific integrated circuit
ASIP	application-specific instruction set processor
ASPP	application-specific programmable processor
ATM	asynchronous transfer mode
BL	bit line in a memory array
CDFG	control data flow graph
CLIW	configurable long instruction word
CMOS	complementary metall oxide semiconductor
COFDM	coded orthogonal frequency division multiplex
COFF	common object file format
CORDIC	coordinate rotation digital computer
COSY	compiler design system (ACE)
CPU	central processing unit
DES	data encryption standard
DFG	data flow graph
DFT	discrete fourier transformation
DMA	direct memory access
DOA	direction-of-arrival (multi-antenna systems)
DSP	digital signal processing or digital signal processor
DSSP	domain specific signal processors
DVB-T	terrestrial digital video broadcasting
E	energy

EDA	electronic design automation
EPIC	explicitly parallel instruction computing
EV	eigenvector or eigenvalue
EVD	eigenvalue/eigenvector decomposition
FFT	fast fourier transformation
FIR	finite impulse response (filter)
FPGA	field-programmable gate array
FSM	finite state machine
FU	functional unit of a processor
GCC	GNU C compiler
GP	general purpose
GPR	general purpose register
GSM	global system for mobile communications
GUI	graphical user interface
HDL	hardware description language
HLL	high level language
HW	hardware
ICORE	ISS-core (the first ASIP designed at the Institute for Integrated Signal Processing Systems)
IDFT	inverse discrete fourier transformation
IIR	infinite impulse response (filter)
ILP	instruction level parallelism
IP	intellectual property
IPO	in-place optimization
IRQ	interrupt request
ISA	instruction set architecture
ISI	inter-symbol interference
ISS	instruction set simulator
JPEG	joint photographic experts group
LISA	language for instruction set architecture description (ISS)
LMS	least mean square
LSB	least significant bit/byte
MIMD	multiple instruction, multiple data (architecture)
MIPS	million instructions per second, commercial processor architecture
MISD	multiple instruction, single data (architecture)

MOS	metall oxide semiconductor
MPEG	motion pictures experts group
MSB	most significant bit/byte
ND2	NAND standard cell with two inputs
NMOS	metall oxide semiconductor with N-field effect transistors
NOP	no-operation instruction
NP	network processor
OFDM	orthogonal frequency division multiplexing
PAST	projection approximation subspace tracking
PC	program counter
PEQ	phase equalization
PSA	programmable system architecture
PTL	processor template library
QAM	quadrature amplitude modulation
RAM	random access memory
RF	register file
RISC	reduced instruction set computer
ROM	read only memory
RTL	register transfer language
SIMD	single instruction, multiple data (architecture)
SISD	single instruction, single data (architecture)
SM	signed magnitude (number representation)
SPARC	scalable processor architecture
SPICE	simulation program with integrated circuit emphasis
SRAM	static random access memory
SVD	singular value decomposition
T	critical path of a synchronous design
TCG	test case generator
VHDL	VHSIC hardware description language
VHSIC	very-high-speed integrated circuits
VLIW	very long instruction word
VLSI	very large scale integration
XOR	exclusive logical OR operation

Bibliography

[1] Arthur Abnous and Jan Rabaey. Ultra-Low-Power Domain-Specific Multimedia Processors. *Proceedings of the VLSI Signal Processing Workshop*, pages 461–470, Oct. 1996.

[2] Santosh Abraham, Bob Rau, Robert Schreiber, Greg Snider, and Michael Schlansker. Efficient Design Space Exploration in PICO. *Proc. of the Conference on Compilers, Architectures and Synthesis for Embedded Systems (CASES)*, pages 71–79, Nov. 2000.

[3] ACE – Associated Compiler Experts bv., http://www.ace.nl. *The COSY Compiler Development System*, 2002.

[4] Douglas Adams. *The Ultimate Hitchhikers's Guide – Complete & unabridged.* Wings Books, 1996.

[5] Advanced Risc Machines Ltd. *ARM7100 Data Sheet*, Dec. 1994.

[6] S. Affes, S. Gazor, and Y. Grenier. An Algorithm for Multi-Source Beamforming and Multi-Target Tracking. *IEEE Trans. on Signal Processing*, 44(6):1512–1522, 1996.

[7] A. Aiken and A. Nicolau. A Realistic Resource-Constrained Software Pipelining Algorithm. *Advances in Languages and Compilers for Parallel Processing, A. Nicola et al., Pitman/The MIT Press, London*, pages 274–290, 1991.

[8] A. Alomary, T. Nakata, Y. Honma, M. Imai, and N. Hikichi. An ASIP Instruction Set Optimization Algorithm with Functional Module Sharing Constraint. *Proc. IEEE/ACM, Int. Conference on Computer Aided Design, Santa Clara, CA, USA*, pages 526–532, Nov. 1993.

[9] G. M. Amdahl. Validity of Single-Processor Approach to Achieving Large-Scale Computing Capability. *Proc. AFIPS Conf. Reston, VA*, pages 483–485, 1967.

[10] Analog Devices Inc. *TigerSHARC Reference Manual*, Dec. 2001.

[11] David P. Appenzeller and Andreas Kuehlmann. Formal Verification of a PowerPC^{TM} Microprocessor, Report RC (19971). Technical report, IBM T. J. Watson Research Center, Yorktown Heights, NY 10598, March 1995.

[12] ARC Cores Ltd. *ARC Programmers Reference Manual*, Dec. 1999.

[13] ARC Cores Ltd. *ARCtangent Processor, http://www.arccores.com*, 2001.

[14] Simon Segars ARM Inc. Low-Power Design Techniques for Microprocessors. In *International Solid State Circuits Conference*, February 4–8, 2001.

[15] ARM Limited, http://www.arm.com. *ARM9E-S Technical Reference Manual*, 1999.

[16] ARM Ltd., Cambridge, UK. *ARM Instruction Set Quick Reference Card, ARM QRC 001D*, 1999.

[17] M. Arnold and Henk Corporaal. Designing Domain-Specific Processors. *Proc. of the 2001 Int. Workshop on Hardware/Software Codesign, Copenhagen, Denmark*, 2001.

[18] Krste Asanovic, Mark Hampton, Ronny Krashinsky, and Emmet Witchel. Energy-Exposed Instruction Sets. *1st Chapter of Power Aware Computing, Editors: Robert Graybill and Rami Melhem, Plenum Publishing, to appear*, 2002.

[19] Semiconductor Industry Association. International Technology Roadmap for Semiconductors. Technical report, http://public.itrs.net/files/1999_SIA_Roadmap/Home.htm, 1999.

[20] Y. Bajot and H. Mehrez. A Macro-Block Based Methodology for ASIP Core Design. In *Proc. of the Int. Conf. on Signal Processing Applications and Technology (ICSPAT)*, Nov. 1999.

[21] S. Balakrishnan and S. K. Nandy. Multithreaded Architectures for Media Processing. *1st Workshop on Media Processors and DSPs (MP-DSP), Haifa, Israel*, Nov. 15, 1999.

[22] M. Barbacci. Instruction Set Processor Specifications (ISPS): The Notation and its Application. *IEEE Transactions on Computers*, C-30(1):24–40, Jan. 1981.

[23] Michael J. Bass and Clayton M. Christinsen. The Future of the Microprocessor Business. *IEEE Spectrum*, pages 34–39, April 2002.

[24] Maurice Bellanger. *Digital Processing of Signals - Theory and Practice*. John Wiley & Sons Ltd., 2000.

[25] L. Benini, G. De Micheli, E. Macii, D. Sciuto, and C. Silvano. Asymptotic Zero-Transition Activity Encoding for Address Busses in Low-Power Microprocessor-Based Systems. *Proc. of the 7th Great Lakes Symposium on VLSI, Urbana, IL*, March 13-15, 1997.

[26] S. B. Bensley and B. Aazhang. Subspace-Based Channel Estimation for Code Division Muliple Access Communication Systems. *IEEE Trans. on Communications*, 1994.

[27] J. Bier. DSP 16xxx Targets Communications Apps. Technical report, Microprocessor Report, Vol. 11, No. 12, Sept. 15, 1997.

[28] A. Bindra. Two 'Lode' up on TCSI's new DSP core. *EE Times*, Jan. 1995.

[29] Nguyen Ngoc Binh, Masaharu Imai, Akichika Shiomi, and Nobuyuki Hikichi. A Hardware/Software Partitioning Algorithm for Designing Pipelined ASIPs with Least Gate Counts. *Proceedings of the Design Automation Conference (DAC 96), Las Vegas, Nevada, USA*, pages 527–532, June 1996.

[30] Matthias A. Blumrich, Kai Li, Richard Alpert, Cezary Dubnicki, and Edward W. Felten. Virtual Memory Mapped Network Interface for the SHRIMP Multicomputer. *Proc. of the 21st Annual International Symposium on Computer Architecture*, pages 142–153, April 1994.

[31] BOPS, Inc., www.bops.com. *Man Array Architecture*, 2000.

[32] G. Boriello. Software Scheduling in the Co-Synthesis of Reactive Real-Time Systems. *Proc. of the 31st Design Automation Conference*, 1994.

[33] Jörg Bormann, Jörg Lohse, Michael Payer, and Gerd Venzl. Model Checking in Industrial Hardware Design. *Proceedings of the Design Automation Conference (DAC95), San Francisco, CA*, June 12–16, 1995.

[34] D.G. Bradlee, R.E. Henry, and S.J. Eggers. The Marion System for Retargetable Instruction Scheduling. In *Proc. of the Int. Conf. on Programming Language Design and Implementation (PLDI)*, pages 229–240, 1991.

[35] G. Braun, A. Hoffmann, A. Nohl, and H. Meyr. Using Static Scheduling Techniques for the Retargeting of High Speed, Compiled Simulators for Embedded Processors from an Abstract Machine Description. In *Proc. of the Int. Symposium on System Synthesis (ISSS)*, Oct. 2001.

[36] Thomas D. Burd and Robert W. Brodersen. Processor Design for Portable Systems. *Journal of VLSI Signal Processing*, 13(2/3):203–222, Aug. 1996.

[37] Thomas D. Burd and Robert. W. Brodersen. Processor Design for Portable Systems. *Journal of VLSI Signal Processing, 13 (2/3)*, pages 203–222, August 1996.

[38] Klaus-Peter Buss and Volker Wittke. Mikro-Chips für Massenmärkte – Innovationsstrategien der europäischen und amerikanischen Halbleiterhersteller in den 90er Jahren. Mitteilungen des Soziologischen Forschungsinstituts (SOFI), Universität Göttingen, July 2000.

[39] Cadence Design Systems, Inc., 2655 Seely Avenue, San Jose, CA 95134, USA. *Cadence Virtual Component Co-Design (VCC)*, 2002.

[40] T. Callaway and E. Swartzlander. Optimizing Arithmetic Elements for Signal Processing. In *IEEE VLSI Signal Processing Workshop*, Oct. 1992.

[41] R. Camposano and J. Wilberg. Embedded System Design. *ACM Transactions on Design Automation for Electronic Systems*, 10(1):5–50, 1996.

[42] Francky Catthoor, Frank Franssen, Sven Wuytack, Lode Nachtergaele, and Hugo De Man. Global Communication and Memory Optimizing Transformations for Low Power Signal Processing Systems. *VLSI Signal Processing Workshop*, pages 178–187, Oct. 1994.

[43] A. Chandra, V. Iyengar, D. Jameson, R. Jawalekar, I. Nair, B. Rosen, M. Mullen, J. Yoon, R. Rmoni, D. Geist, and Y. Wolfsthal. AVPGEN – A Test Generator for Architecture Verification. *IEEE Transactions on Very Large Scale Integration (VLSI) Systems*, 3(2):188–199, 1995.

[44] A. Chandrakasan and R. Brodersen. Minimizing Power Consumption in Digital CMOS Circuits. In *Proc. of the IEEE*, volume 83,4, April 1995.

[45] A. Chandrakasan, M. Potkonjak, R. Mehra, J. Rabaey, and R. Brodersen. Optimizing Power Using Transformations. In *IEEE Transactions on Computer Aided Design*, volume 14,1, pages 12–31, Jan. 1995.

[46] A. P. Chandrakasan and R. W. Brodersen. *Low Power Digital CMOS Design*. Kluwer Academic Publishers, 1995.

[47] Chang. Power-Area Tradeoff in Divided Word Line Memory Arrays. *Journal of Circuits, Systems, Computers*, 1(7):49–57, 1997.

[48] D. G. Chinnery, B. Nikolic, and K. Keutzer. Achieving 550 MHz in an ASIC Methodology. *DAC2001*, June 18-22, Las Vegas, Nevada, USA 2001.

[49] H. Choi, I.C. Park, S.H. Hwang, and C.M. Kyung. Synthesis of Application Specific Instructions for Embedded DSP Software. In *Proc. of the Int. Conf. on Computer Aided Design (ICCAD)*, Nov. 1998.

[50] H. Choi, J.H. Yi, J.Y. Kee, I.C. Park, and C.M. Kyung. Exploiting Intellectual Properties in ASIP Designs for Embedded DSP Software. In *Proc. of the Design Automation Conference (DAC)*, Jun. 1999.

[51] D. Chuang, G. Kamosa, D. Arya, and R. Priebe. JPEG2000: A Scalable and Configurable Multiprocessor VLIW Implementation. *Mobile Multimedia Conference 2000, Tokyo, Japan*, Oct. 2000.

[52] A. Cilio. Efficient Code Generation for ASIPs with Different Word Sizes. *Proc. of the 3rd Conference of the Advanced School for Computing and Imaging, The Netherlands*, June 1997.

[53] Andrea G. M. Cilio and Henk Corporaal. Global Program Optimization: Register Allocation of Static Scalar Objects. *Proc. of the 5th Annual Conference of the advanced School for Computing and Imaging, Delft*, pages 52–57, June 1999. ISBN: 90-803086-4-1.

[54] Theo A. C. M. Claasen. High Speed: Not the Only Way to Exploit the Intrinsic Computational Power of Silicon. *Proc. of the 1999 IEEE International Solid-State Circuits Conference*, pages 22–25, 1999.

[55] I. Bernhard Cohen. *Howard Aiken: Portrait of a Computer Pioneer*. MIT Press, March 1999.

[56] Robert Cohn and Mark Vandevoorde. Instrumentation, and Profile Based Optimization. In *Tutorial 4 at the International Conference on Architectural Support for Programming Languages and Operating Systems, San Jose, CA*, http://www.tru64unix.compaq.com/dcpi/publications.htm, 1998.

[57] J. W. Cooley and J. W. Tukey. An Algorithm for the Machine Computation of Complex Fourier Series. *Math. Comp.*, 19:297–301, 1965.

[58] Joel Mc Cormack, Robert Mc Namara, Christopher Gianos, Larry Seiler, Norman P. Jouppi, Ken Correll, Todd Dutton, and John Zurawski. WRL Research Report 98/1: Neon - A (Big) (Fast) Single-Chip 3D Workstation Graphics Accelerator. Technical report, COMPAQ, Western Research Laboratory, Palo Alto, CA, USA, 1998.

[59] Thomas H. Cormen, Charles E. Leiserson, and Ronald L. Rivest. *Introduction to Algorithms*. MIT Press, ISBN 0-262-03141-8, 1989.

[60] Olivier Coudet. The Quest for Timing Closure. In *Tutorial 3, Part 4, ASP-DAC 2001, Yokohama, Japan*, http://cadlab.cd.ucla.edu/ cong/slides/asp-dac01 tutorial/t3 p4 cou.pdf, 2001.

[61] CoWare. *http://www.coware.com.*

[62] H. Dawid and H. Meyr. The Differential CORDIC Algorithm: Constant Scale Factor Redundant Implementation without correcting Iterations. *IEEE Transactions on Computers*, 45(3):307–318, March 1996.

[63] J. Demmel and K. Veselic. Jacobi's Method is more accurate than QR. Technical report, Courant Institute of Mathematical Sciences, Department of Computer Science, New York University, 1989.

[64] S. Devadas, A. Ghosh, and K. Keutzer. *Logic Synthesis*. Mc Graw Hill, NY, 1994.

[65] S. Devadas, A. Ghosh, and K. Keutzer. An Observability-Based Code Coverage Metric for Functional Simulation. *Proc. of the International Conference on Computer-Aided Design*, pages 418–425, 1996.

[66] Srinivas Devadas and Sharad Malik. A Survey of Optimization Techniques Targeting Low Power VLSI Circuits. *32nd Design Automation Conference*, 1995.

[67] S. Dey, Y. Gefen, A.C. Parker, and M. Potkonjak. *Wiley Encyclopedia of Electrical and Electronics Engineering: Critical Path Analysis and Minimization in Digital Circuits*, volume 4, pages 404–415. John Wiley & Sons, 2000.

[68] Nathan Dohm, Carl Ramey, Darren Brown, Scot Hildebrandt, James Huggins, Mike Quinn, and Scott Taylor. Zen and the Art of Alpha Verification. *Int. Conference on Computer Design (ICCD98), Austin, Texas*, Oct. 5–7, 1998.

[69] F. Engel, J. Nührenberg, and G.P. Fettweis. A Generic Tool Set for Application Specific Processor Architectures. In *Proc. of the Int. Workshop on Hardware/Software Codesign*, Apr. 1999.

[70] J.-H. Yang et al. Metacore: An Application Specific DSP Development System. In *Proc. of the Design Automation Conference (DAC)*, Jun. 1998.

[71] K. Usami et al. Design Methodology of Ultra Low-Power MPEG4 Codec Core Exploiting Voltage Scaling Techniques. In *Proceedings of the 35th Design Automation Conference, San Francisco, CA, USA*, pages 483–488, 1998.

[72] S. C. Huang et al. RLC Signal Integrity Analysis of High-Speed Global Interconnects. In *International Electron Devices Meeting (IEDM), Tech. Digest*, 2000.

[73] S. Gary et al. The PowerPC 603 Microprocessor: A Low-Power Design for Portable Applications. In *Proceedings of the 39th IEEE Computer Society International Conference*, pages 307–315, March 1994.

[74] S. Kobayashi et al. Compiler Generation in PEAS-III: an ASIP Development System. In *Proc. of the Workshop on Software and Compilers for Embedded Systems (SCOPES)*, Mar. 2001.

[75] European Telecommunication Standards Institute, Sophia Antipolis, France. *ETSI ETS 300744, Digital Video Broadcasting: Framing Structure, Channel Coding and Modulation for Digital Terrestrial Television, V1.1.2*, 1997.

[76] P. Faraboschi, G. Desoli, and J.A. Fisher. Very Long Instruction Word Architectures for DSP and Multimedia Applications: The Latest Word in Digital and Media Processing. *IEEE Signal Processing Magazine*, pages 59–85, Mar. 1998.

[77] Paolo Faraboschi, Geoffrey Brown, Joseph A. Fisher, Giuseppe Desoli, and Fred Homewood. Lx: A Technology Platform for Customizable VLIW Embedded Processing. *Proc. 27th Annual International Symposium on Computer Architecture (ISCA-2000)*, pages 203–213, June 12–14, 2000.

[78] A. Fauth, M. Freericks, and A. Knoll. Generation of Hardware Machine Models from Instruction Set Descriptions. In *Proc. of the IEEE Workshop on VLSI Signal Processing*, 1993.

[79] A. Fauth and A. Knoll. Automatic Generation of DSP Program Development Tools Using a Machine Description Formalism. In *Proc. of the Int. Conf. on Acoustics, Speech and Signal Processing (ICASSP)*, 1993.

[80] Jay Fenlason and Richard Stallman. *GNU gprof - The GNU Profiler*. Free Software Foundation, Inc., 1997.

[81] Gerhard Fettweis, M. Weiss, W. Drescher, U. Walther, F. Engel, S. Kobayaschi, and T. Richter. Breaking New Grounds Over 3000M MAC/s: A Broadband Mobile Multimedia Modem DSP. *Proc. of the Int. Conf. on Signal Processing Applications and Technology (ICSPAT'98), Toronto Canada*, pages 1547–1551, 1998.

[82] M. J. Flynn. Some Computer Organizations and their Effectiveness. *IEEE Trans. Computers*, 21(9):948–960, September 1972.

[83] W. Fornaciari, D. Sciuto, and C. Silvano. Power Estimation for Architectural Exploration of HW/SW Communication on System-Level Buses. In *Proc. of the Int. Workshop on Hardware/Software Codesign (CODES99)*, Rome, Italy, May 3–5, 1999.

[84] M. Freericks. The nML Machine Description Formalism. Technical Report, Technical University of Berlin, Department of Computer Science, 1993.

[85] C. Ghez, M. Miranda, A. Vandecappelle, F. Catthoor, and D. Verkest. Systematic High-Level Address Code Transformations for Piece-Wise Linear Indexing Illustrations: Illustration on a Medical Algorithm. In *Proceedings IEEE Workshop on Signal Processing Systems, Lafayette, LA, USA*, Oct. 2000.

[86] T. Glökler, Stefan Bitterlich, and Heinrich Meyr. Power Reduction for ASIPs: A Case Study. *IEEE Workshop on Signal Processing Systems (SIPS), Antwerpen, Belgium*, 2001.

[87] Tilman Glökler, Stefan Bitterlich, and Heinrich Meyr. DSP Core Verification Using Automatic Test Case Generation. *IEEE Trans. Acoust., Speech and Signal Processing*, June 2000.

[88] Tilman Glökler, Stefan Bitterlich, and Heinrich Meyr. ICORE: A Low-Power Application Specific Instruction Set Processor for DVB-T Acquisition and Tracking. *13th IEEE workshop on Signal Processing Systems (ASIC/SOC'2000)*, September 2000.

[89] Tilman Glökler, Stefan Bitterlich, and Heinrich Meyr. Increasing the Power Efficiency of Application Specific Instruction Set Processors using Datapath Optimization. *Workshop on Sig. Proc. Syst. SIPS2000*, October 2000.

[90] Tilman Glökler, Stefan Bitterlich, and Heinrich Meyr. Power-Efficient Semi-Automatic Instruction Encoding for Application Specific Instruction Set Processors. *IEEE Trans. Acoust., Speech and Signal Processing*, May 2001.

[91] Tilman Glökler, Andreas Hoffmann, and Heinrich Meyr. Methodical Low-Power ASIP Design Space Exploration. *Kluwer Journal of VLSI Signal Processing (JVSP)*, Volume 33, Issue 3:229–246, March 2003.

[92] Tilman Glökler and Heinrich Meyr. ASIP Design and the Energy-Flexibility Tradeoff. In *Proceedings of the 10th Aachen Symposium on Signal Theory, ISBN 3-8007-2610-6, Aachen, Germany*, pages 343–348, September 2001.

[93] Steve Golson. Resistance is Futile! Building Better Wireload Models. In *Synopsys Users Group Conference (SNUG99), San Jose*, 1999.

[94] G. H. Golub and C. F. Van Loan. *Matrix Computations*. North Oxford Academic, 1989.

[95] J. Gong, D.D. Gajski, and S. Narayan. Software Estimation Using a Generic-Processor Model. In *Proc. of the European Design and Test Conference (ED&TC)*, Mar. 1995.

[96] Jie Gong, Daniel D. Gajski, and Alex Nicolau. A Performance Evaluator for Parameterized ASIC Architectures. *Proc. of the European Conference on Design Automation (EDAC)*, pages 66–71, 1994.

[97] R. Gonzales. Xtensa: A Configurable and Extensible Processor. *IEEE Micro*, 20(2):60–70, Mar. 2000.

[98] G. Goosens, J. van Praet, D. Lanneer, W. Geurts, A. Kifli, C. Liem, and P. G. Paulin. Embedded Software in Real-Time Signal Processing Systems: Design Technologies. *IEEE Proc. Spec. Issue on Hardware/Software Codesign*, 1996.

[99] R. Govindarajan, Erik R. Altman, and Guang R. Gao. A Theory for Software-Hardware Co-Scheduling for ASIPs and Embedded Processors. *Proc. of the Int. Conf. on Application Specific Systems, Architectures, and Processors (ASAP), Boston, MA, USA*, July 10–12, 2000.

[100] B. Grattan, G. Stitt, and F. Vahid. Codesign-Extended Applications. *IEEE/ACM International Symposium on Hardware/Software Codesign, Estes Park*, May 2002.

[101] M. Gschwind. Instruction Set Selection for ASIP Design. In *Proc. of the Int. Workshop on Hardware/Software Codesign*, May 1999.

[102] A. Halambi, P. Grun, V. Ganesh, A. Khare, N. Dutt, and A. Nicolau. EXPRESSION: A Language for Architecture Exploration through Compiler/Simulator Retargetability. In *Proc. of the Conference on Design, Automation & Test in Europe (DATE)*, Mar. 1999.

[103] S. Hanono, G. Hadjiyiannis, and S. Devadas. Aviv: A Retargetable Code Generator Using ISDL. *Tech Report, SPAM Project, Princeton University*, April 1996.

[104] Paul J. Havinga and Gerard J. M. Smit. Design Techniques for Low Power Systems. *Journal of System Architectures*, 46(1):1–21, 2000. ISSN: 1383-7621.

[105] S. Haykin. *Adaptive Filter Theory.* Prentice Hall, 1991.

[106] Gerben J. Hekstra and Ed F. Deprettere. A Chip Set For A Ray-Casting Engine. *Proc. of the Workshop on VLSI Signal Processing IX, San Francisco, CA, USA*, 1996.

[107] J. Hennessy and D. Patterson. *Computer Architecture: A Quantitative Approach.* Morgan Kaufmann Publishers Inc., 1996. Second Edition.

[108] Richard C. Ho, C. Han Yang, Mark A. Horowitz, and David L. Dill. Architecture Validation for Processors. *Proc. of the International Symposium on Computer Architecture (ISCA)*, pages 404–413, 1995.

[109] A. Hoffmann, T. Kogel, A. Nohl, G. Braun, O. Schliebusch, A. Wieferink, and H. Meyr. A Novel Methodology for the Design of Application Specific Instruction Set Processors (ASIP) Using a Machine Description Language. *IEEE Transactions on Computer-Aided Design*, 20(11):1338–1354, Nov. 2001.

[110] A. Hoffmann, H. Meyr, and R. Leupers. *Architecture Exploration for Embedded Processors with LISA.* Kluwer Academic Publishers, 2002.

[111] A. Hoffmann, A. Nohl, G. Braun, and H. Meyr. Generating Production Quality Software Development Tools Using A Machine Description Language. In *Proc. of the Conference on Design, Automation & Test in Europe (DATE)*, Mar. 2001.

[112] A. Hoffmann, O. Schliebusch, A. Nohl, G. Braun, O. Wahlen, and H. Meyr. A Methodology for the Design of Application Specific Instruction-Set Processors Using the Machine Description Language LISA. In *Proc. of the Int. Conf. on Computer Aided Design (ICCAD)*, Nov. 2001.

[113] B. Holmer and B. Prangle. Hardware/Software Codesign Using Automated Instruction Set Design & Processor Synthesis, 1993.

[114] Mark Horowitz. Lecture 24: Power, Low Power Design. *http://www.cs.utexas.edu/users/skeckler/cs384v/ handouts/lecture24_2.pdf, Stanford University*, 1999.

[115] I.J. Huang and A.M. Despain. Generating Instruction Sets and Microarchitectures from Applications. In *Proc. of the Int. Conf. on Computer Aided Design (ICCAD)*, Nov. 1994.

[116] I.J. Huang and A.M. Despain. Synthesis of Instruction Sets for Pipelined Microprocessors. In *Proc. of the Design Automation Conference (DAC)*, Jun. 1994.

[117] I.J. Huang, B. Holmer, and A.M. Despain. ASIA: Automatic Synthesis of Instruction-Set Architectures. In *Proc. of the SASIMI Workshop*, Oct. 1993.

[118] G. A. Van Huben. The Role of Two-Cycle Simulation in the S/390 Verification Process. *IBM Journal of Research & Development*, 41(4/5), 1997.

[119] David A. Huffman. A Method for the Construction of Minimum-Redundancy Codes. *Proc. of the IRE*, 40(9):1098–1101, 1952.

[120] Kai Hwang and Zhiwei Xu. *Scalable Parallel Computing.* McGraw-Hill International Editions, 1997.

[121] IBS. Analysis of SOC Design Costs, A Custom Study for Synopsys Professional Services. Technical report, International Business Strategies, Inc., 632 Industrial Way, Los Gatos, CA 95030, USA, February 2002.

[122] Berkeley Design Technology Inc. VLIW Architectures for DSP: A Two Part Lecture. *Proceedings of ICSPAT99, the International Conference on Signal Processing Applications and Technology*, November 1–4, 1999.

[123] INFINEON Technologies AG, Munich, Germany. *Product Brief SQC 6100 - Terrestrial Receiver for DVB-T*, 2000. www.infineon.com/products/ics/pdf/sqc_10b.pdf.

[124] M. J. Irwin. Low Power Design for Systems on a Chip (SOCs). Tutorial at the 12th Annual IEEE ASIC/SOC Conference, September 1999.

[125] Sergio Akira Ito, Luigi Carro, and Ricardo Pezzuol Jacobi. System Design Based on Single Language and Single-Chip Java ASIP Microcontroller. *Proc. of the Conference on Design, Automation & Test in Europe (DATE), Paris, France*, pages 703–707, March 2000.

[126] M. Itoh, S. Higaki, J. Sato, A. Shiomi, Y. Takeuchi A. Kitajima, and M. Imai. PEAS-III: An ASIP Design Environment. In *Proc. of the Int. Conf. on Computer Design (ICCD)*, Sep. 2000.

[127] M. Itoh, Y. Takeuchi, M. Imai, and A. Shiomi. Synthesizable HDL Generation for Pipelined Processors from a Micro-Operation Description. *IEICE Transactions on Fundamentals of Electronics, Communications and Computer Sciences*, E83-A(3), Mar. 2000.

[128] C. G. J. Jacobi. Über ein leichtes Verfahren die in der Theorie der Säcularstörungen vorkommenden Gleichungen numerisch aufzulösen. *Crelle's Journal*, 30, 1846.

[129] Carsten Jacobi. Evaluation and Implementation of DSP Core Architectures for COFDM. Master's thesis, Supervisor: T. Glökler, D 418, Institute for Integrated Signal Processing Systems (ISS), RWTH Aachen, Jul. 2000.

[130] Margarida F. Jacome and Gustavo de Veciana. Lower Bound On Latency For VLIW ASIP Datapath. *Proc. of the Int. Conf. on Computer Aided Design (ICCAD99)*, pages 261–269, November 1999.

[131] Manoj Kumar Jain, Lars Wehmeyer, Peter Marwedel, and M. Balakrishnan. Register File Synthesis in ASIP Design. *Tech. Report No. 747, University of Dortmund, Dept. of CSXII*, 2001.

[132] M.K. Jain, M. Balakrishnan, and A. Kumar. ASIP Design Methodologies: Survey and Issues. In *Int. Conf. on VLSI Design*, Jan. 2001.

[133] Curtis L. Janssen. *VPROF - Programmer's Manual*, 2002.

[134] Poul M. Rands Jensen. On Jacobi-Like Algorithms for Computing the Ordinary Singular Value Decomposition. Technical Report R 91 - 33, Department of Communication Technology, Institute for Electronic Systems, Aalborg University, Denmark, Oct. 1991.

[135] Robert B. Jones. Efficient Validity Checking for Processor Verification. *IEEE International Conference on Computer Aided Design (ICCAD)*, 1995.

[136] Andrew B. Kahng. Design Technology Productivity in the DSM Era. *Invited Talk to the Asia South-Pacific Design Automation Conference, Yokohama*, February 2001.

[137] Vinod Kathail, Michael S. Schlansker, and B. Ramakrishna Rau. HPL-PD Architecture Specification: Version 1.1. Technical report, Compiler and Architecture Research, HP Laboratories Palo Alto, Feb. 2000.

[138] Stephan Keil. Untersuchung von Prozessorarchitekturen und generische VHDL-Implementierung eines skalierbaren digitalen Signalprozessors. Master's thesis, Supervisor: T. Glökler, D 411Institute for Integrated Signal Processing Systems (ISS), RWTH Aachen, Dec. 1999.

[139] T. M. Kemp, R. K. Montoye, J.D. Harper, J.D. Palmer, and D. J. Auerbach. A Decompression Core for PowerPC. *IBM Journal of Research and Development*, 42(6):807–812, Nov. 1998.

[140] Bart Kienhuis, Ed Deprettere, Kees Vissers, and Pieter van der Wolf. The Construction of a Retargetable Simulator for an Architecture Template. *International Workshop on Hardware/Software Co-design, IEEE Computer Society, CODES/CASHE'98, Seattle Washington, USA*, March 15–18, 1998.

[141] P. Kievits, E. Lambers, C. Moerman, and R. Woudsma. R.E.A.L. DSP Technology for Telecom Baseband Processing. *Proc. of the Int. Conf. on Signal Processing Applications and Technology (ICSPAT), Toronto, CA*, 1998.

[142] K.-W. Kim, T. T. Hwang, C. L. Liu, and S.-M. Kang. Logic Transformation for Low Power Synthesis. In *Design, Automation and Test in Europe Conference and Exhibition*, Munich, Germany, March 9–12, 1999.

[143] Kyosun Kim, Ramesh Karri, and Miodrag Potkonjak. Synthesis of Application Specific Programmable Processors. *Proc. of the Design Automation Conference (DAC 97), Anaheim, California*, 1997.

[144] Y.-W. Kim, Y.-M. Yang, J.-T. Yoo, and S.-W. Kim. Low-Power Digital Filtering Using Approximate Processing with Variable Canonic Signed Digit Coefficients. In *IEEE Int. Symposium on Circuits and Systems*, May 28–31, 2000.

[145] Y.G. Kim and T.G. Kim. A Design and Tool Reuse Methodology for Rapid Prototyping of Application Specific Instruction Set Processors. In *In Proc. of the Workshop on Rapid System Prototyping (RSP)*, Apr. 1999.

[146] A. Kitajima, M. Itoh, J. Sato, A. Shiomi, Y. Takeuchi, and M. Imai. Effectiveness of the ASIP Design System PEAS-III in Design of Pipelined Processors. In *Proc. of the Asia South Pacific Design Automation Conference (ASPDAC)*, Jan. 2001.

[147] S. Klauke and J. Götze. Low Power Algorithms for Signal Processing. *ITG Workshop Mikroelektronik für die Informationstechnik, Darmstadt, Germany*, 2000.

[148] Peter Voigt Knudsen and Jan Madsen. Aspects of Modelling in Hardware/Software Partitioning. *Int. Workshop on Rapid System Prototyping, Thessaloniki, Greece*, June 19–21, 1996.

[149] Gerd Krüger. A Tool for Hierarchical Test Generation. *IEEE Transactions on Computer-Aided Design*, 10(4):519–524, April 1991.

[150] Kayan Kücükcakar. An ASIP Design Methodology for Embedded Systems. *7th International Workshop on HW/SW Co-Design, Roma, Italy*, 1999.

[151] R. Kumaresan and D. W. Tufts. Estimating the Angles of Arrival of Multiple Plane Waves. *IEEE Trans. on Aerospace Electron. Syst.*, AES-19:134–139, 1983.

[152] M. Kuulusa, J. Nurmi, J. Takala, P. Ojala, and H. Herranen. A Flexible DSP Core for Embedded Systems. *IEEE Design & Test of Computers*, 14(4):60–68, 1997.

[153] Young-Jun Kwon, Danny Parker, and Hyuk Jae Lee. TOE: Instruction Set Architecture for Code Size Reduction and Two Operations Execution. *Int. Workshop on Compiler and Architecture Support for Embedded Systems, Washington D. C., USA*, Oct. 1–3, 1999.

[154] L. R. Rabiner and B. G. Gold. *Theory and Application of Digital Signal Processing*. Englewood Cliffs, NJ: Prentice-Hall, 1975.

[155] M. Lam. Software Pipelining: An Effective Scheduling Technique For VLIW Machines. *Proc. of the SIGPLAN'88 Conf. On Programming Language and Implementation, Atlanta, GA*, pages 318–328, June 1988.

[156] D. Lanner, J. Van Praet, A. Kifli, K. Schoofs, W. Geurts, F. Thoen, and G. Goossens. Chess: Retargetable Code Generation for Embedded DSP Processors. In P. Marwedel and G. Goosens, editors, *Code Generation for Embedded Processors*. Kluwer Academic Publishers, 1995.

[157] Hsien-Hsin Lee, Youfeng Wu, and Gary Tyson. Accurate Invalidation Profiling for Effective Data Speculation on EPIC Processors. *International Symposium on Performance Analysis of Systems and Software*, Apr. 8–10, 2000.

[158] Lea Hwang Lee, Bill Moyer, and John Arends. Instruction Fetch Energy Reduction Using Loop Caches for Embedded Applications with Small Tight Loops. *IEEE Symp. on Low Power Electronics and Design, San Diego, CA*, Aug. 16–17, 1999.

[159] Lea Hwang Lee, Bill Moyer, and John Arends. Instruction Fetch Energy Reduction Using Loop Caches For Embedded Applications with Small Tight Loops. *IEEE Int. Symposium on Low Power Electronics and Design, San Diego, CA*, August 16–17, 1999.

[160] S.H. Leibson. Jazz Joins VLIW Juggernaut – CMP and Java as an HDL Take System-on-Chip Design to Parallel Universe. *Microprocessor Report*, 2000.

[161] C. E. Leierson, F. M. Rose, and J. B. Saxe. Optimizing Synchronous Circuitry by Retiming. *Proc. 3rd Caltech Conference on VLSI*, pages 23–36, March 1883.

[162] Jeroen A. J. Leijten, Jef L. van Meerbergen, Adwin H. Timmer, and Jochen A. G. Jess. PROPHID: A Data-Driven Multi-Processor Architecture for High-Performance DSP. *Proceedings of the 1997 European Design and Test Conference (ED&TC'97)*, 1997.

[163] H. Lekatsas, J. Henkel, and W. Wolf. Code Compression for Low Power Embedded System Design. *Int. Workshop on Hardware/Software Co-Design*, 2000.

[164] H. Lekatsas and W. Wolf. Random Access Decompression using Binary Arithmetic Coding. *Proc. of the 1999 IEEE Data Compression Conference*, March 1999.

[165] Rainer Leupers and Peter Marwedel. Retargetable Generation of Code Selectors from HDL Processor Models. *European Design and Test Conference*, pages 140–144, 1997.

[166] Rainer Leupers and Peter Marwedel. Retargetable Code Generation Based on Structural Processor Descriptions. *Design Automation for Embedded Systems*, 3, Jan. 1998.

[167] Jeremy Levitt and Kunle Olukotun. A Scalable Formal Verification Methodology for Pipelined Microprocessors. *Proceedings of the Design Automation Conference (DAC 96), Las Vegas, Nevada*, 1996.

[168] Daniel Lewin, Dean Lorenz, and Shmuel Ur. A Methodology for Processor Implementation Verification. *First Int. Conf. on Formal Methods in Computer Aided Design, Springer Verlag*, 1166:126–142, 1996.

[169] Y.-T. S. Li, S. Malik, and A. Wolfe. Performance Estimation for Embedded Software with Instruction Cache Modeling. *Int. Conf. on Computer-Aided Design (ICCAD)*, pages 380–387, 1995.

[170] Clifford Liem, Trevor May, and Pierre Paulin. Register Assignment through Resource Classification for ASIP Microcode Generation. *Proc. of the Int. Conference on Computer Aided Design (ICCAD)*, Nov. 1994.

[171] Markus Lorenz, Rainer Leupers, and Peter Marwedel. Low-Energy DSP Code Generation Using a Genetic Algorithm. *Int. Conf. on Computer Design (ICCD), Austin, Texas*, Sept. 2001.

[172] J. Ludwig, S. Nawab, and A. Chandrakasan. Low-Power Digital Filtering Using Approximate Processing. *IEEE Journal of Solid-State Circuits, Vol. 31, No. 3*, 31(3), March 1996.

[173] F. Maessen, A. Giulietti, B. Bougard, V. Derudder, L. Van der Perre, F. Catthoor, and M.Engels. Memory Power Reduction for the High-Speed Implementation of Turbo Codes. In *IEEE Workshop on Signal Processing Systems (SIPS)*, 2001.

[174] Steven T. Mangelsdorf, Raymond P. Gratias, Richard M. Blumberg, and Rohit Bhatia. Functional Verification of the HP PA 8000 Processor. *Hewlett-Packard Journal, Article 3*, Aug. 1997.

[175] Peter Marwedel, Stefan Steinke, and Lars Wehmeyer. Compilation Techniques for Energy-, Code-Size-, and Run-Time-Efficient Embedded Software. *Workshop on Advanced Compiler Techniques for High Performance and Embedded Processors, Bucharest*, 2001.

[176] Jean-Marc Masgonty, Stefan Cserveny, and Christian Piguet. Low-power sram and rom memories. *International Workshop-Power And Timing Modeling, Optimization and Simulation (PATMOS 2001), Yverdon-Les-Bains, Switzerland*, September 26–28, 2001.

[177] Mahesh Mehendale, Sunil D. Sherlekar, and G. Venkatesh. Low-Power Realization of FIR Filters on Programmable DSPs. *IEEE Transactions on Very Large Scale Integration (VLSI) Systems*, 6(4), Dec. 1998.

[178] H. Meyr and G. Ascheid. *Synchronization in Digital Communications*, volume 1. John Wiley & Sons, 1990.

[179] H. Meyr, M. Moeneclaey, and S. A. Fechtel. *Digital Communication Receivers*. John Wiley & Sons, Inc., Wiley Series in Telecommunication and Signal Processing edition, 1998.

[180] MIPS Technologies Inc., Mountain View, CA. *MIPS32 4Kc Processor Core Datasheet*, 1999.

[181] M. Miranda, C. Ghez, C. Kulkarni, F. Catthoor, and D. Verkest. Systematic Speed-Power Memory Data-Layout Explorations for Cache Controlled Embedded Multimedia Applications. *Proceedings of the International Symposium of System Synthesis (ISSS), Montreal, Quebec*, Oct. 1–3, 2001.

[182] F. Monssen. *MicroSim PSpice with Circuit Analysis, 2nd Edition*. Prentice Hall, Upper Saddle River, NJ, 1998.

[183] J. Monteiro, S. Devadas, and A. Ghosh. Retiming Sequential Circuits for Low Power. *Proc. IEEE International Conference on Computer Aided Design (ICCAD)*, pages 398–402, Nov. 1993.

[184] G. Moore. Cramming more Components onto Integrated Circuits. *Electronics magazine*, Apr. 1965.

[185] V. Moshnyaga. Adaptive Bit-Width Compression for Low Energy Frame Memory Design. In *IEEE Workshop on Signal Processing Systems (SIPS)*, 2001.

[186] Motorola. *SC110 DSP Core Reference Manual*, Apr. 2001.

[187] Motorola Inc. *M•Core Reference Manual.*

[188] G. J. Myers. *The Art of Software Testing*. John Wiley & Sons, New York, 1976.

[189] Chetana Nagendra, Robert Michael Owen, and Mary Jane Irwin. Low Power Tradeoffs in Signal Processing Hardware Primitives. In J. Rabaey and P. M. Chau, editors, *VLSI Signal Processing*, volume VII, pages 276–285. IEEE Signal Processing Society Press, 1994.

[190] Chetana Nagendra and Robert Michael Owens. Power-Delay Characteristics of CMOS Adders. *IEEE Transactions on VLSI Systems*, 2(3):377–381, Sept. 1994.

[191] F. Najm, I. Haji, and P. Yang. Electromigration Median Time-to-Failure based on Stochastic Current Waveform. In *Proc. of the IEEE International Conference on Computer Design*, Nov. 1989.

[192] J. Von Neumann. First draft of a report on the EDVAC. In N. Stern, editor, *From ENIAC to Univac: An Appraisal of the Eckert-Mauchly Computer*. Digital Press, Bedford, Massachusetts, 1945.

[193] A. Nohl, G. Braun, O. Schliebusch, A. Hoffmann, R. Leupers, and H. Meyr. A Universal Technique for Fast and Flexible Instruction-Set Architecture Simulation. In *Proc. of the Design Automation Conference (DAC)*, 2002.

[194] Tobias G. Noll. The Deep-Submicron Nightmare, Invited Speaker. In *Workshop on Embedded Systems and Applications, Livigno, Italy*, March 1996.

[195] F. Onion, A. Nicolau, and N. Dutt. Incorporating Compiler Feedback into the Design of ASIPs. *Proceedings of the 1995 European Design and Test Conference (ED&TC'95)*, 1995.

[196] Keshab K. Parhi. Fast Low-Energy VLSI Binary Addition. *International Conference on Computer Design*, pages 676–684, 1997.

[197] Keshab K. Parhi. *VLSI Digital Signal Processing Systems*. John Wiley & Sons, 1999.

[198] David Y. W. Park, Jens U. Skakkebaek, Mats P. E. Heimdahl, Barbara J. Czerny, and David L. Dill. Checking Properties of Safety Critical Specifications Using Efficient Decision Procedures. *Proc. of the 2nd Workshop on Formal Methods in Software Practice (FMSP'98), St. Petersburg, FL*, March 1998.

[199] P. Paulin. Design Automation Challenges for Application-Specific Architecture Platforms. Keynote speech at SCOPES 2001 - Workshop on Software and Compilers for Embedded Systems (SCOPES), Apr. 2001.

[200] P. Paulin, F. Karim, and P. Bromley. Network Processors: A Perspective on Market Requirements, Processor Architectures and Embedded SW Tools. In *Proc. of the Conference on Design, Automation & Test in Europe (DATE)*, Mar. 2001.

[201] P. Paulin, C. Liem, C. May, and S. Sutarwala. CodeSyn: A Retargetable Code Synthesis System. In *Proc. of the Int. Symposium on System Synthesis (ISSS)*, May 1994.

[202] P. Paulin, C. Liem, T.C. May, and S. Sutarwala. FlexWare: A Flexible Firmware Development Environment for Embedded Systems. In *Code Generation for Embedded Processors, Editors: P. Marwedel and G. Goossens*. Kluwer Academic Publishers, 1995.

[203] A. Paulraj, R. Roy, and T. Kailath. A Subspace Approach to Signal Parameter Estimation. *Proceedings of the IEEE, 74:1044-1045*, 1986.

[204] Craig Peterson, Tim Elliott, and Naveed Sherwani. Seven Critical Scaling Challenges of ASIC Design. *http://www.intel.com/design/asics/pdf/Micro-ASIC_wp.pdf*, 2001.

[205] Philips, http://www.semiconductor.philips.com/acrobat/literature/9397/75007159.pdf. *Trimedia Data Book*, 2001.

[206] Padmanabhan Pillai and Kang G. Shin. Real-Time Dynamic Voltage Scaling for Low-Power Embedded Operating Systems. In *18th ACM Symposium on Operating Systems Principles*, 2001.

[207] P. Plöger and J. Wildberg. A Design Example Using CASTLE. *Workshop on Design Methodology for Microelectronics, Inst. for Computer Systems, Slovak Academy of Science, Bratislava, Slovakia*, pages 160–167, Sept. 1995.

[208] J. Van Praet, D. Lanneer, G. Goossens, W. Geurts, and H. De Man. A Graph Based Processor Model for Retargetable Code Generation. *Proc. of the European Design & Test Conference*, March 1996.

[209] S. Ramanathan, V. Visvanathan, and S. K. Nandy. Synthesis of ASIPs for DSP Algorithms. *Integration, The VLSI Journal*, June 1999.

[210] B. Ramakrishna Rau. Iterative Modulo Scheduling. *International Journal of Parallel Processing*, 24(1), Feb. 1996.

[211] B.R. Rau and M.S. Schlansker. Embedded Computer Architecture and Automation. *IEEE Computer*, 34(4):75–83, Apr. 2001.

[212] John D. Ruley. The Future of Moore's Law, Part 2. *BYTE Magazine*, June 2001.

[213] Y. Saad and M. Schultz. Data Communication in Parallel Architectures. *Journal of Parallel and Distributed Computing*, Feb. 1989.

[214] H. Savoj, R. Brayton, and H. Touati. Extracting Local Don't Care for Network Optimization. *Proc. of the Int. Conf. on Computer Aided Design (ICCAD)*, pages 514–517, 1991.

[215] Lou Scheffer. Timing Closure Today. In *Tutorial 3, Part 2, ASP-DAC 2001, Yokohama, Japan*, http://cadlab.cd.ucla.edu/ cong/slides/asp-dac01_tutorial/t3_p2_lou.pdf, 2001.

[216] U. Schlichtmann. Design Reuse: Experiences at Siemens Semiconductor and Future Directions. *http://www.ecsi.org/ecsi/Doc/OtherDoc/IPreuse/PDF/siemens.pdf*, Oct. 10, 1997.

[217] O. Schliebusch, A. Hoffmann, A. Nohl, G. Braun, and H. Meyr. Architecture Implementation Using the Machine Description Language LISA. In *Proc. of the ASPDAC/VLSI Design - Bangalore, India*, Jan. 2002.

[218] Oliver Schliebusch, Andreas Hoffmann, Achim Nohl, Gunnar Braun, and Heinrich Meyr. Architecture implementation using the machine description language lisa. *ASP-DAC/VLSI Design 2002, Bangalore, India*, pages 239–244, Jan. 2002.

[219] R. O. Schmidt. *A Signal Subspace Approach to Multiple Emitter Location and Spectral Estimation*. PhD thesis, Standord University, Nov. 1981.

[220] B. Schneier. *Applied Cryptography (2nd Edition). Protocols, Algorithms, and Source Code in C*. John Wiley & Sons, 1996.

[221] Pieter J. Schoenmakers and J. Frans M. Theeuwen. Clock Gating on RT-Level VHDL. *Proc. of the Int. Workshop on Logic Synthesis, Tahoe City, CA*, pages 387–391, June 7–10, 1998.

[222] Robert Schreiber, Shail Aditya, Scott Mahlke, and Vinod Kathail. PICO-NPA: High-Level Synthesis of Nonprogrammable Hardware Accelerators. Technical report, Hewlett-Packard Company, Laboratories, Palo Alto, CA 94304, 2001.

[223] Jeff Scott, Lea Hwang Lee, John Arends, and Bill Moyer. Designing the Low Power M•Core Architecture. *Power Driven Microarchitecture Workshop at the IEEE Int. Symposium on Circuits and Systems (ISCAS), Barcelona, Spain*, June 1998.

[224] Jeff Scott, Lea Hwang Lee, Bill Moyer, and John Arends. Assembly-Level Optimizations for the M•Core^{TM} M3 Processor Core. *Int. Workshop on Compiler and Architecture Support for Embedded Systems*, Oct. 1999.

[225] B. Shackleford, M. Yasuda, E. Okushi, H. Koizumi, H. Tomiyama, and H. Yasuura. Satsuki: An Integrated Processor Synthesis and Compiler Generation System. In *IEICE Transactions on Information and Systems*, pages 1373–1381, 1996.

[226] C. E. Shannon. A Mathematical Theory of Communications. *The Bell System Technical Journal, Vol. 27, pp. 379-423, pp. 623-656*, Jul./Oct. 1948.

[227] Deszso Sima, Terence Fountain, and Péter Kacsuk. *Advanced Computer Architectures - A Design Space Approach*. Addison-Wesley, 1997.

[228] *The SPARC Architecture Manual Version V8*, 1992.

[229] R.M. Stallman. *Using and Porting the GNU Compiler Collection*. Free Software Foundation, gcc-2.95 edition, 1999.

[230] Stefan Steinke, Rüdiger Schwarz, Lars Wehmeyer, and Peter Marwedel. Low Power Code Generation for a RISC Processor by Register Pipelining. *Technical Report #754, Department of Computer Science, University of Dortmund*, 2001.

[231] Greg Stitt, Frank Vahid, Tony Givargis, and Roman Lysecky. A First Step Towards an Architecture Tuning Methodology for Low Power. *Int. Conf. on Compilers, Architectures, and Synthesis for Embedded Systems (CASES), San Jose, CA*, Nov. 2000.

[232] R. Sucher, R. Niggebaum, G. Fettweiss, and A. Rom. CARMEL - A New High Performance DSP Core using CLIW. *http;//www.carmledsp.com*, 1999.

[233] Synopsys. *DesignWare Foundation Library Databook, Volume 1-3, Release 2001.08, August 2001.*

[234] Synopsys. *CoCentric System Studio*
http://www.synopsys.com/products/cocentric_studio/cocentric_studio.html, 2001.

[235] Synopsys, Inc. *Design Compiler User Guide, Version 2001.08, August 2001.*

[236] Synopsys, Inc., http://www.synopsys.com. *Module Compiler User Guide, Version 2001.08, August 2001.*

[237] Synopsys, Inc., 700 East Middlefield Road, Mountain View, CA 94043, USA. *Formality User's Guide, Version 2001.06-FM1*, 2001.

[238] Synopsys, Inc., 700 East Middlefield Road, Mountain View, CA 94043, USA. *CoCentric System Studio Data Sheet*, 2002.

[239] G. Martin S. Swan T. Grötker, S. Liao. *System Design with SystemC*. Kluwer Academic Publishers, 2002.

[240] Target Compiler Technologies. *CHESS/CHECKERS/BRIDGE/DARTS/GO*
http://www.retarget.com, 2001.

[241] Serdar Tasiran and Kurt Keutzer. Coverage Metrics for Functional Verification of Hardware Designs. *IEEE Design & Test of Computers*, pages 2–11, July/August 2001.

[242] J. Teich, R. Weper, D. Fischer, and S. Trinkert. BUILDABONG: A Rapid Prototyping Environment for ASIPs. In *Proc. of the DSP Germany (DSPD)*, Oct. 2000.

[243] Tensilica. *Xtensa*
http://www.tensilica.com, 2001.

[244] Texas Instruments. *TMS320C2x User's Guide*, Jan. 1993.

[245] Texas Instruments. *TMS320C54x CPU and Instruction Set Reference Guide*, Oct. 1996.

[246] Texas Instruments. *TMS320C62x/C67x CPU and Instruction Set Reference Guide*, Mar. 1998.

[247] Adwin H. Timmer, Marino T. J. Srik, Jef L. Van Meerbergen, and Jochen A. G. Jess. Conflict Modelling and Instruction Scheduling in Code Generation for In-House DSP Cores. *Proc. of the 32nd Design Automation Conference, San Francisco, CA*, pages 593–598, 1995.

[248] V. Tiwari, P. Ashar, and S.Malik. Technology Mapping for Low Power. In *Proceedings of the Design Automation Conference*, pages 74–79, 1993.

[249] V. Tiwari, D. Singh, S. Rajgopal, G. Mehta, R. Patel, and F. Baez. Reducing Power in High-Performance Microprocessors. *Design Automation Conference (DAC98)*, pages 732–737, 1998.

[250] Vivek Tiwari and Mike Tien-Chien Lee. Power Analysis of a 32-bit Embedded Microcontroller. *VLSI Design Journal*, 7(3), 1998.

[251] Vivek Tiwari, Sharad Malik, and Pranav Ashar. Guarded Evaluation: Pushing Power Managment to Logic Synthesis/Design. In *International Symposium on Low Power Design, Dana Point, CA*, April 1995.

[252] Vivek Tiwari, Sharad Malik, and Andrew Wolfe. Compilation Techniques for Low Energy: An Overview. *Symposium on Low-Power Electronics, San Diego, CA*, Oct. 1994.

[253] Vivek Tiwari, Sharad Malik, and Andrew Wolfe. Instruction Level Analysis and Optimization of Software. *Journal of VLSI Signal Processing*, 13(2), 1996.

[254] Hiroyuki Tomiyama, Tohrue Ishihara, Akihiko Inoue, and Hiroto Yasuura. Instruction Scheduling for Power Reduction in Processor-Based System Design. *Proc. of the Conference on Design, Automation & Test in Europe (DATE)*, Feb. 1998.

[255] Trimaran. http://www.trimaran.org, 2001. An Infrastructure for Research in Instruction-Level Parallelism.

[256] Barbara Tuck. Linking Logical and Physical Design. *Electronic Systems*, 38(1):58–64, 1999.

[257] A. Turier, L. Ben Ammar, and A. Amara. An Accurate Power and Timing Modeling Technique Applied To A Low-Power ROM Compiler. *Proc. of the Int. Workshop on Power, and Timing Modeling, Optimization and Simulation (PATMOS), Lyngby, Technical University of Denmark*, Oct. 7–9, 1998.

[258] A. Tyagi. Energy-Time Trade-offs in VLSI Computation. In *Proc. of the 9th Conference on the Foundations of Software Technology and Theoretical Computer Science, LNCS series*. Springer Verlag, 1989.

[259] Jul. 1994 United States Environmental Protection Agency, EPA 430-K-94-006. Purchasing An Energy Star Computer.

[260] Shmuel Ur and Yoav Yadin. Micro Architecture Coverage Directed Generation of Test Programs. *Proceedings of the Design Automation Conference (DAC99), New Orleans, LA*, pages 175–180, June 1999.

[261] Cary Ussery. Configurable Processing Platforms: Redefining SoC. *http://www.improvsys.com*, 2001.

[262] J. van Praet, G. Goossens, D. Lanner, and H. De Man. Instruction Set Definition and Instruction Selection for ASIPs. In *Proc. of the Int. Symposium on System Synthesis (ISSS)*, Oct. 1994.

[263] Martin Vaupel. *Effizienter Entwurf eines DVB-Satelliten-Empfängers*. PhD thesis, Institute for Integrated Signal Processing Systems, RWTH Aachen, Shaker Verlag, ISBN 3-8265-6676-9, 1999.

[264] Luis Villa, Michael Zhang, and Krste Asanovic. Dyamic Zero Compression for Cache Energy Reduction. *33rd Int. Symp. on Microarchitecture, Monterey, CA*, Dec. 2000.

[265] S. Virtanen, J. Lilius, and T. Westerlund. A Processor Architecture for the TACO Protocol Processor Platform. *Proceedings of the 18th IEEE NorChip Conference, Turku, Finland*, Nov. 6–7, 2000.

[266] J.E. Volder. The CORDIC Trigonometric Computing Technique. *IRE Transactions on Electronic Computers*, Volume EC-8(no.3):330–334, Sept. 1959.

[267] O. Wahlen, T. Glökler, A. Nohl, A. Hoffmann, R. Leupers, and H. Meyr. Application Specific Compiler/Architecture Codesign: A Case Study. In *Proc. of the Joint Conference on Languages, Compilers, and Tools for Embedded Systems (LCTES) and Software and Compilers for Embedded Systems (SCOPES)*, Jun. 2002.

[268] O. Wahlen, M. Hohenauer, R. Leupers, and H. Meyr. Using Virtual Resources for Generating Instruction Schedulers. In *Proc. of IEEE/ACM International Workshop on Application Specific Processors (WASP'02)*, Nov. 2002.

[269] O. Wahlen, M. Hohenauer, R. Leupers, and H. Meyr. Instruction scheduler generation for retargetable compilation. In *IEEE Design & Test of Computers*, Jan. 2003.

[270] O. Wahlen, M.Hohenauer, G. Braun, R. Leupers, G. Ascheid, H. Meyr, and Xiaoningg Nie. Extraction of Efficient Instruction Schedulers from Cycle-True Processor Models. In *Proc. of the Conference on Sofware and Compilers for Embedded Systems (SCOPES)*, Sep. 2003.

[271] Oliver Wahlen and Tilman Glökler et al. Application Specific Architecture/Compiler Codesign: A Case Study. *SCOPES Workshop 2002, Berlin, Germany*, June 19–21, 2002.

[272] Marlene Wan, Yuji Ichikawa, David Lidsky, and Jan Rabaey. An Energy Conscious Methodology for Early Design Exploration of Heterogeneous DSPs. *Proc. of the Custom Integrated Circuits Conference, Santa Clara, CA*, May 1998.

[273] C.-Y. Wang and K. Roy. An Activity-Driven Encoding Scheme for Power Optimization in Microprogrammed Control Unit. *IEEE Transactions on Very Large Scale Integration (VLSI) Systems*, 7(1):130–134, March 1999.

[274] A. Wieferink, T. Kogel, A. Nohl, A. Hoffmann R. Leupers, and H. Meyr. A Generic Toolset for SoC Multiprocessor Debugging and Synchronisation. In *"IEEE International Conference on Application-Specific Systems, Architectures and Processors (ASAP)"*, June 2003.

[275] M. Willems, H. Keding, and V. Živojnović. Modulo-Addressing Utilization in Automatic Software Synthesis for Digital Signal Processors. In *Proceedings of the IEEE International Conference on Acoustics, Speech and Signal Processing (ICASSP)*, pages 287–290, München, Apr. 1997.

[276] M. Willems and V. Živojnović. DSP-Compiler: Product Quality for Control-Dominated Applications? In *Proc. of the Int. Conf. on Signal Processing Applications and Technology (ICSPAT)*, Oct. 1996.

[277] Phillip J. Windley. Formal Modeling and Verification of Microprocessors. *IEEE Transactions on Computers*, 44(1):54–72, 1995.

[278] A. Wolfe and A. Chanin. Executing Compressed Programs on an Embedded RISC Architecture. *Proc. 25th Annual International Symposium on Micro-Architectures, Portland, OR, USA*, pages 81–91, Dec. 1992.

[279] B. Yang. Projection Approximation Subspace Tracking. *IEEE Trans. on Signal Processing*, 43(1):95–107, Jan. 1995.

[280] J.-H. Yang, B.-W. Kim, S.-J. Nam, Y.-S. Kwon, D.-H. Lee, J.-Y. Lee, C.-S. Hwang, Y.-H. Lee, S.-H. Hwang, I.-C. Park, and C.-M. Kyung. MetaCore: An Application-Specific Programmable DSP Development System. *IEEE Transactions on Very Large Scale Integration (VLSI) Systems*, 8(2):173–183, Apr. 2000.

[281] Gary Yeap. *Practical Low Power Digital VLSI Design*. Kluwer Academic Publishers, 1998.

[282] Qin Zhao, Bart Mesmann, and Twan Basten. Practical Instruction Set Design and Compiler Retargetability Using Static Resource Models. *Proceedings of the IEEE Design Automation and Test in Europe, DATE 2002, Paris, France*, March 2002.

[283] V. A. Zivkovic, R. J. W. T. Tangelder, and H. G. Kerkhoff. Design and Test Space Exploration of Transport-Triggered Architectures. *Proc. of the Conference on Design, Automation & Test in Europe (DATE), Paris, France*, pages 146–151, March 27, 2000.

[284] V. Živojnović, S. Pees, Ch. Schläger, M. Willems, R. Schoenen, and H. Meyr. DSP Processor/Compiler Co-Design: A Quantitative Approach. In *Proc. of the IEEE Symposium on System Synthesis - La Jolla*, Nov. 1996.

[285] V. Živojnović, H. Schraut, M. Willems, and R. Schoenen. DSPs, GPPs, and Multimedia Applications: An Evaluation using DSPstone. In *Proc. Int. Conf. on Signal Processing Application and Technology (ICSPAT)*, Boston, Oct. 1995.

[286] V. Živojnović, S. Tjiang, and H. Meyr. Compiled Simulation of Programmable DSP Architectures. In *Proc. of the IEEE Workshop on VLSI Signal Processing*, Oct. 1995.